Hydroponics for Beginners:

A Simple Introduction and Guide to Gardening Without Using Soil

William Garcia

Hydroponics for Beginners

© Copyright 2020 by William Garcia. All right reserved.

The work contained herein has been produced with the intent to provide relevant knowledge and information on the topic on the topic described in the title for entertainment purposes only. While the author has gone to every extent to furnish up to date and true information, no claims can be made as to its accuracy or validity as the author has made no claims to be an expert on this topic. Notwithstanding, the reader is asked to do their own research and consult any subject matter experts they deem necessary to ensure the quality and accuracy of the material presented herein.

This statement is legally binding as deemed by the Committee of Publishers Association and the American Bar Association for the territory of the United States. Other jurisdictions may apply their own legal statutes.

Any reproduction, transmission or copying of this material contained in this work without the express written consent of the copyright holder shall be deemed as a copyright violation as per the current legislation in force on the date of publishing and subsequent time thereafter. All additional works derived from this material may be claimed by the holder of this copyright.

The data, depictions, events, descriptions and all other information forthwith are considered to be true, fair and accurate unless the work is expressly described as a work of fiction. Regardless of the nature of this work, the Publisher is exempt from any responsibility of actions taken by the reader in conjunction with this work. The Publisher acknowledges that the reader acts of their own accord and releases the author and Publisher of any responsibility for the observance of tips, advice, counsel, strategies and techniques that may be offered in this volume.

Table of Contents

INTRODUCTION 7

CHAPTER 1: WHAT IS HYDROPONICS? 9

- THE PAST 9
- ADVANCEMENTS IN 17TH AND 18TH CENTURY 12
- HYDROPONICS AND ITS PRINCIPLES 16
- HOW DOES THE SYSTEM OF HYDROPONICS WORK? 18
- COMPONENTS OF HYDROPONICS SYSTEM 19
- THE ROOT STORY 21
- MEDIUMS OF HYDROPONIC 23

CHAPTER 2: ADVANTAGES AND DISADVANTAGES OF HYDROPONICS 31

- ADVANTAGES 31
- DISADVANTAGES 38
- IS HYDROPONIC SYSTEM RECOMMENDED? 42

CHAPTER 3: TYPES OF HYDROPONICS 44

- DEEP WATER CULTURE SYSTEMS 44
- WICK SYSTEM 46
- NUTRIENT FILM TECHNIQUE 49
- EBB AND FLOW HYDROPONIC SYSTEM 51
- DRIP HYDROPONIC SYSTEM 54
- AEROPONICS 56

CHAPTER 4: FEATURES OF A HYDROPONIC GARDEN 59

- WATER PUMPS 59
- AIR STONES AND AIR PUMPS 61
- RESERVOIRS, FLOOD TABLES AND TRAYS 63
- LIGHTING SYSTEM 65

CHAPTER 5: HOW TO SET UP A HYDROPONICS SYSTEM 69

- HOW TO CHOOSE THE PERFECT LIGHT? 69
- DIFFERENT PLANTS AND DIFFERENT NEEDS OF LIGHT 70
- PARTS OF LIGHTING SYSTEM NECESSARY FOR HYDROPONIC SETUP 71

NUTRIENTS AS THE CORE OF A HYDROPONIC SYSTEM 73
16 PLANT NUTRIENTS 74
MEASURING NUTRIENTS USING EC 77
NUTRIENT FORMULAS AND RATIOS 78
MIXING UP SOLUTIONS 79
SETTING UP HYDROPONIC GARDEN 79

CHAPTER 6: PLANTS SUITABLE FOR THE HYDROPONIC GARDEN 85

LETTUCE 85
TOMATO 86
KALE 87
BEANS 88
RADISH 89
CUCUMBER 90
SPINACH 90
BROCCOLI 91
STRAWBERRIES 91
BLUEBERRIES 92

CHAPTER 7: PLANT CULTURING 95

SEEDING 95
PRODUCTION OF SEEDLING 96
CULTURE OF TOMATO SEEDLING 97
CULTURE OF CUCUMBER SEEDLING 99
CULTURE OF PEPPER SEEDLING 101
CULTURE OF EGGPLANT SEEDLING 102
CULTURE OF LETTUCE SEEDLING 102

CHAPTER 8: PEST CONTROL 104

COMMON TYPES OF PEST PROBLEMS 104
AVOIDING THINGS THAT MIGHT INVITE IN PESTS INTO THE GROWING AREA 106
HOW CAN YOU IDENTIFY PEST PROBLEMS? 109
THINGS TO DO WHEN YOUR SYSTEM HAS PEST INFESTATION 111
INFORMATION ON PESTS AND TREATMENTS 113

CHAPTER 9: EXTRA TIPS AND TRICKS FOR BEGINNERS 117

- Planning the Location 118
- Choosing the Plant Types That You Want to Grow 118
- Making a Proper Plan 119
- Choosing the Ideal System of Hydroponics 120
- Take Care of the Lighting Conditions 120
- Maintaining Proper Temperature 121
- Germination of Seeds 122
- Planting Process 122
- Checking Water Quality 122
- Selecting Proper Nutrients 123
- Checking Health of Plant Roots From Time to Time 123
- Making the Required Investment 124

CHAPTER 10: COMMON MISTAKES AND HOW TO AVOID THEM....125

- Underestimating the System Building Costs 126
- Selecting the Wrong Type of Crops 126
- Not Paying Attention to the pH Levels 127
- Using Wrong Nutrients 128
- Overwatering 129
- Insufficient Lighting 130

CONCLUSION 131

Introduction

Congratulations on purchasing *Hydroponics for Beginners: A Simple Introduction And Guide To Gardening Without Using Soil,* and thank you for doing so.

The majority of people love to do gardening in their house. But, what stops them from doing so is not having enough space. This is very much applicable for all those people who stay in apartments or crowded cities where there is a lack of space. Also, the climate of the place where you are living in is also a big factor as a proper climate is needed for suitable conditions of growth. Then comes hydroponics with which you can make your dream of gardening turn into reality by using some super easy tactics. Hydroponics is a special way of growing plants where there is no need for soil.

In hydroponics, the plant roots are anchored by using a growth medium. For the proper growth of the plants, the required nutrient solutions are also provided. Also, this method is much faster than the traditional way of growing plants as you will need to provide everything to the plants that they need only within a required level and that too in an excellently controlled way. You will be learning everything along with the steps that you need to know about hydroponics in this book.

There are plenty of books on this subject on the market, thanks again for choosing this one! Every effort was made to ensure it is full of as much useful information as possible. Enjoy!

Chapter 1: What Is Hydroponics?

Hydroponics is the modern technology of growing plants. Although it might seem like a new system to many, it is one of the oldest systems of plant growth along with the ocean-growing algae and bacteria photosynthetic bacteria that actually existed much before the terrestrial form of plants, helping in producing an oxygenated environment in which we all breathe today. The modern-day system of hydroponics was actually developed from some of the findings of experiments that were out for determining the composition of plants that dated back to the early days of 1600. But, much before that, plants were already being grown in the culture where soil is not used, although it was not at all identified as being hydroponics directly.

The Past

The first case of hydroponics that was based on the water was in Babylon in the Hanging Gardens, which is among the Seven Wonders of the World. The gardens used to live on an elaborate system of watering which supplied a constant stream of water from the river which was rich in minerals and oxygen. Located right on the eastern banks of the river named Euphrates which is close to the present-day Baghdad, the gardens were being built up by King Nebuchadnezzar II in order to please Amyitis, his wife (604 – 562 BC).

Similar to this, the hieroglyphics of the ancient Egypt that dates back to hundreds of years BC also depicts growing plants right along the river Nile without the use of soil, as with the Chinese floating gardens, that was also described by the great traveler Marco Polo in his very popular journal.

In Central America, the Aztecs also developed a very ingenious way of using up the principles of hydroponics. While they were being treated with sheer hostility by the powerful neighbors and as they were being deprived of any form of arable land, they learned the technique of building rafts by using reeds and rushes that they called chinampas. Chinampas were tough roots and stalks that were being lashed together and were loaded up with the sediments from the shallow bottom of the lake. As the sediment actually originated right from the bottom of the lake, they were loaded up with minerals and organic compounds that were being used up by the Aztecs for the purpose of growing and nourishing plants. The chinampas were able to support the abundant growth of flowers and vegetables, and even large trees. The plant roots grew across the floor of chinampas that allowed a constant source of water along with oxygenation of roots.

The chinampas were also joined together at times for forming islands somewhat around 200 feet long that were being flanked by the waterways and also the drainage canals. Some of the chinampas even consisted of huts for the gardeners. The chinampas turned out to be a great success, and they even supported a civilization of more than 2.000.000 people at the time of the Aztec rule, thus making it much larger than other cities in Europe of that time. A simple makeshift village that was actually invented merely out of desperation of creative type for staving off various demise finally resulted into an established horticulture system that was easily capable of supporting the entire capitol city of that time Central Mexico, a successful testament to the great efficiency of future no soil culture.

As the Spaniards came to the New World, the very sight of the floating islands over water made Cortes and his whole gang astonished. The historian, William Prescott, who wrote about the Aztec empire destruction by the invading Spaniards, explained about the chinampas as the islands of Verdure where vegetables and fruits were seen moving just like rafts over the river water. The use of Chinampas continued very well into the modern 19th century. Somewhat similar procedure also flourished in present-day Bolivia, Peru, and

Ecuador much before the arrival of Columbus in the brand new world. Some of the functional experiments about the system can still be seen today in Xochimilco in the city of Mexico and Tlaxcala state.

The oldest records of scientific approach for discovering the needs of growing plants dates back to 1600 when the scientist Belgian Jan Helmont depicted in his experiment that plants also take in substances from the water. He planted a willow shoot of 5 pounds in a small tube that contained about 200 pounds of soil, dried, isolated for ensuring the accuracy. After a time period of five years, with regular watering of the shoot by using rainwater, Belgian found out that the weight of the willow shoot increased by almost 150 pounds, whereas the weight of the soil was lost by almost two ounces. Therefore, his assumption of plants obtaining substances from the water for its growth was actually correct. But, he failed to conclude that plants also need oxygen and carbon dioxide right from the air.

Advancements In 17th and 18th Century

The modern aspect of chemistry made huge progress during the time period of the 17th and 18th centuries. In addition to the revolution in the methods of science, advancements were also made in the aspect of scientific research. The capability of all the scientists for working off an overall agreed platform of chemically based compounds resulted in a more presumptuous and spirited debate on the basic nutritional needs for the proper growth of plants and also laid the basic foundation of the present-day perception of requirements for growth of plants.

In the year 1792, the scientist Joseph Priestly found out that plants, when placed in a closed chamber containing carbon dioxide, will start absorbing it gradually and will be giving off oxygen. After some years, Jean Housz also carried out the work of Joseph Priestly and took the same one step ahead by discovering that when plants are placed in a chamber containing carbon dioxide, it can easily replace the carbon dioxide with oxygen only within few hours when the chamber is placed in direct sunlight.

Housz established that the process tends to work even faster when bright light is added to the step, and only the green plant parts are involved. With the help of several experiments during the time period of the mid 19th century, the scientists started determining the plant composition and also the substances that are needed for its growth. It was found out that soil was not much important directly for the growth of plants other than just holding the roots and the minerals in place for growth. Although this was only a form of generalization of all the dynamic and critical role of microbes and humus for the health of soil systems, it actually created a new road for greater understanding of the requirements of plant growth and also the technology of hydroponic.

In place of the soil itself, it was found that it was actually the minerals that were mixed up in the soil along with the spaces in between them, that was all that the plants actually thrived for. The very next step that was set up for the articulation of the technology related to hydroponics was to discard the medium of growth and then grow the plants in a solution of water that included all the minerals that were necessary for plant growth. In the year 1860, Professor Julius Sachs of the botany department of the Wurzburg University came up with the first formula of standard nature for the solution of nutrients that can be mixed up with water and in which plant growth was possible.

The technique came to be known as nutriculture.

All these investigations about the nutrients requirements of plants demonstrated that for the normal plant growth, it is possible to grow the same by immersing the plant roots in a solution of water that contains Phosphorus (P), Nitrogen (N), Calcium (Ca), Potassium (K), Sulfur (S), and Magnesium (Mg). The other elements- oxygen, carbon, and hydrogen are derived from the air and water. All these nine elements came to be known as macronutrients. With further progress in the techniques of science, seven other elements were also found out that the plants need in small quantities for their growth, known as the trace elements or micronutrients. The trace elements included Chlorine (Cl), Iron (Fe), Boron (B), Manganese (Mn), Zinc (Zn), Molybdenum (Mo), and Copper (Cu).

The total number of elements that are required for the growth of plants is actually debatable, where some say it is 17, and others say it is 15. The practical application in the development of nutriculture was not seen until the year 1925 when the industry of greenhouse showed their interest in the usage of the same. The soils of a greenhouse are bound to be replaced at regular intervals of time for overcoming several problems related to soil structure, pests, and fertility. All of these problems were removed by the soilless culture. Dr. Victor Tiedjens is often regarded as the earliest pioneers of the soilless cultivation during the 1920s and 1930s. He found out that plants can only absorb nutrients of fertilizer when they are in the liquid state. He bypassed soil and started applying liquid fertilizer on the plants directly.

In the year 1929, Dr. William Gericke of California University transformed his laboratory of nutriculture into the operation of the commercial form of crop production. He called his systems of nutriculture 'hydroponic' in which hydro means water and Ponos in Greek means working. The government started sponsoring the experiments from the year 1939 when the second World War started.

Right after World War II, the military started using the systems of hydroponics as the only method of food production overseas. With the adoption of plastic in hydroponics during the 1970s, a viable medium of cultivation was found. Plastics freed the growers from the continuous construction as well as the destruction of properties for the early set of components. With timely development of timers, pumps, plastic plumbing, and also effective growing media, systems of hydroponics can now be computerized, automated, and also streamlined for the reduction in the capital along with costs of operation. It is now being adopted in the main cultivation of plants and is widely used for organic form of cultivation. The only scarcity that is being predicted by scientists is water and food scarcity that might result in a big issue during the 21st century.

Hydroponics and Its Principles

Hydroponics is the process of plant cultivation without the use of soil. Hydroponic herbs, flowers, and vegetables are all grown after planting an inert media of growth and are supplied with solutions that are rich in nutrients along with water and oxygen. The system of hydroponics facilitates the rapid growth of plants, superior quality, and also stronger yields. As you grow a plant in soil, the roots of the plant will look out for all its necessary nutrients for supporting its own life. When the roots of a plant are directly exposed to nutrition and water, the plant will no longer need to spend any extra energy for sustaining its own life. All the energy that would have been spent by the plant for acquiring water and food can now be directed for its own maturation. So, as a result, the growth of leaves also tends to flourish along with the blooming of flowers and fruits.

Plants tend to sustain their lives by the process of photosynthesis. They capture the light of the sun by the use of chlorophyll, a green plant pigment present in plant leaves. The energy of light is being used up by the plants for splitting up the molecules of water that has been absorbed by them through the system of roots. The molecules of hydrogen combine up with the available carbon oxide for producing carbohydrates that are being used up by the plants for their own nourishment. After that, oxygen is released out in the atmosphere, which is often regarded as a very crucial factor for preserving the habitability of the planet.

Most of us think that plants cannot sustain without the availability of soil, but in actual, plants do not have any requirement of soil for photosynthesizing. The main need for soil is only to provide the plants with nutrients and water. When nutrients are effectively dissolved in water, they can be directly applied to the root system of the plants by immersion, misting, or flooding. The innovations in the system of hydroponics have provided that with direct exposure of nutrient-rich water, the growth of the plants will be more effective and also versatile when compared with the traditional system of irrigation.

How Does the System Of Hydroponics Work?

The system of hydroponics work by permitting minute control right over the conditions of the environment such as pH balance and temperature and also maximum exposure to water and required nutrients. It operates by using a principle that is very simple in nature: providing the plants with exactly what they require at the time when they need the same. The system of hydroponics administers solutions of nutrients that are tailored for the requirements of any specific plant that is being grown. It will allow you to properly control the amount of light that the plants will receive and also the time behind the same. PH levels can also be adjusted and monitored effectively. In a highly controlled as well as customized type of environment, there is acceleration in the growth of plants.

By controlling the plant environment, you will be able to reduce various risk factors. All those plants that are grown in fields and gardens are, most of the time, introduced to a variety of variables that can negatively impact the growth and health of the plants. Even the fungus that is present in the soil can result in diseases for the plants. Wildlife such as rabbits can destroy the ripening fruits and veggies from the garden. Also, when plants are grown in fields and gardens, they are exposed to pests like locusts that can destroy the growth of the plants. The system of hydroponics can easily cut out the unpredictability of the plant growth outdoors. Without any sort of resistance of a mechanical type of the soil, the seedlings can easily mature and also at a faster rate. With the elimination of pesticides, hydroponics can provide you with high quality and healthy veggies and fruits. Without any sort of obstacle, the plants can grow rapidly and vigorously.

Components of Hydroponics System

For maintaining a flourishing system of hydroponics, you will need certain components for running the system efficiently.

Media of Growth

The plants of the hydroponics system are generally grown in an inert type of media that can easily support the weight of the plant and can support the structure of the root. Growing media is a substitute in place of soil. But, it does not provide the soil with any sort of independent nutrition. The porous nature of growing media can effectively retain all the nutrients and moisture from the solution of nutrients that are applied to the media. The same is then delivered to the plant. There are various growing media that are pH neutral by nature, as well. So, the media will not be upsetting the balance of the nutrient solution. You can choose from a wide range of growing media. The type of the plant and also the system of hydroponic will be dictating the type of media that will suit you the best. The growing media for hydroponics systems are available all over the internet and also at the local stores.

Air Pumps and Air Stones

All those plants that are submerged in the water might drown quickly if the provided water is not aerated sufficiently. Air stones can disperse very tiny bubbles of dissolved oxygen all throughout the reservoir of nutrient solution. These oxygen bubbles also help in even distribution of the nutrients that are dissolved in the solution. You need to note that air stones are not capable of generating oxygen on their own. They are needed to be directly attached to some sort of air pump externally with the help of opaque plastic tubes that are food grade. You can easily get air stones and air pumps in the local store as they are very popular components of the aquarium.

Net Pots

These are mesh type planters that are used for holding hydroponic plants. The latticed kind of material permits the plant roots to grow from the sides, and the pot bottom provides more exposure to nutrients and oxygen. They can also provide excellent drainage than any other plastic of clay pots.

The Root Story

The system of roots varies in structure and size from those of a small seedling, which is only of few inches, to that of a huge tree that generally comes along with an extensive system of roots. No matter what is the size of the roots, there are three main functions that are served by roots:

- Uptakes nutrients and water

- Helps in storing the manufactured materials

- Provides a physical form of support to the plant

In the case of hydroponics, everything revolves around roots. The process of water and nutrient absorption takes place through the root hairs that are present right behind the tip of the root. The root hairs are of extremely delicate nature and generally die off when the root tip extends into the growth medium. Roots absorb nutrients and water through the process of diffusion. As a system of hydroponics is much cleaner than any organic environment of growing, it can very easily provide for a superior quality of growing environment. But, you are required to note that the very first principle on which hydroponics works is GIGO: garbage in and garbage out.

While talking about plant roots, oxygen might not into at the very first stage. But, in actuality, it is very important for the proper health of roots. Oxygen gets absorbed by the plant roots and is then used up by the plant for its own growth. When there is an absence of oxygen near the zone of the root, it might lead to asphyxiation, which might even damage the plant roots and will be affecting the health of the plant. Also, water stagnation near the roots can also lead to asphyxiation.

Mediums of Hydroponic

Growing media are the materials on which the plants grow in a hydroponic system. In traditional cultivation, soil is used as the medium. But, in the case of hydroponics, soil is kept out of the list as it is all about soilless cultivation. That is the main reason why right before finding a suitable hydroponics system, you are required to find out the perfect medium for your hydroponic cultivation.

Importance of Growing Medium

The growing medium will be helping the plant roots by providing the same with oxygen and nutrients that they need. It also plays a deep part in supporting the weight of the plant. Another important role of growing media is that it permits the roots of a plant to have maximum kind of exposure to the necessary nutrients. You will need to moisten the medium of growth with a solution of nutrients. When you use growing media instead of soil, you can have several benefits over the traditional system of cultivation. There will be lesser chances of developing pests and soil-borne diseases. This will allow you to grow high-quality and healthy plants.

You must have heard of growing media such as gravel, sand, peat moss, vermiculite, and perlite. But, there are several types of growing media that are available all around us. So, choosing the perfect medium might turn out to be a daunting task, and it is very important as well.

What Are the Conditions for a Perfect Medium?

A perfect medium will be the one that:

- Is organic in nature, environment friendly, and biodegradable

- Can keep an even proportion of water to air

- Is easy to find and affordable

- Helps in protecting plants from changing pH

- Is light in weight

Now, let's have a look at some of the popular growing mediums.

Perlite

It is a very common medium that is used for hydroponics cultivation and has been around for several years. It is used by the traditional gardeners as it helps in adding aeration to the soil. It is a type of volcanic glass and is mined for extraction. It is created under conditions of extreme heat. It is very porous in nature and is also light in weight. This is a great choice for the wick type system of hydroponics, which we will be discussing in the later chapters. But, because of the porous nature, it can flow out very easily. So, it is only recommended for the small-sized and moderate systems of watering as it might get washed off very easily. It comes with the capability of holding air very well and also has a neutral level of pH. Only because of its easy flowing nature, it is generally used by mixing with some other kind of medium. It is reusable and inexpensive as well.

Coconut Coir

Coconut coir, also known as coco peat, is a byproduct that you can get from coconut. It is made from the brown-colored husks that surround the shell of a coconut. What makes this medium a great choice as a medium is that it is completely organic in nature and is inert as well. It can hold up water very easily. It also comes with a great ratio of air to water, and thus, it can save the growing plants from drowning. It is renewable in nature and is also environmentally friendly. It can either be used alone as a medium or mixed up with perlite. But, the only drawback of this medium is that does not come with great drainage capabilities and also gets uncompressed after using a few times.

Vermiculite

It is a form of laminar minerals that is hydrated and resembles mica. Similar to perlite, this medium is also processed after exposing the same to high heat for expanding them into clean, small, and odorless granules. It acts as a great medium for soilless culture. It is sterile, non-toxic, resistant to moist, and also comes along with a neutral level of pH. It is light in weight and can also hold up water very easily. But, it cannot keep up with aeration very well. As it can hold up water, there are high chances of suffocating the plants. So, for normal usage, it is generally mixed up with other media as well.

Rockwool

This material is being used up in hydroponics cultivation for a long time, especially in the large-sized farms. It is sterile in nature and is also porous. It is made up of limestone or granite rocks, heated up for melting, and then spun into long and thin fibers. The fibers are then compressed into cubes or brick-like structure. It comes with immunity to microbes and can also retain air and water very well. Thus they are great for protecting the plants from dehydration while providing the roots with proper availability of oxygen. But, the natural pH level of this medium is quite high, and thus, it might alter the pH level of the nutrient solution. You can prevent this by soaking the rockwool bricks in the water right before using it. Also, they are non-degradable in nature. The unused fibers cannot be disposed of easily. Additionally, the dust from this medium can also result in skin irritation. So, keeping in mind the safety of the environment, this medium is not being used up much today.

Leca

Leca is like small balls with a similar size as of a marble. This medium is created after heating and then expanding clay for forming small balls of a bubble shape. This is an effective growing media that is being used for hydroponic cultivation. They are light in weight but can provide the plants with all the support that they need. They are also porous in nature and can also properly maintain the balance between water and oxygen. It comes with a neutral level of pH and is reusable as well. In fact, you can clean them sterilize them right before using them again. But, they cannot hold up water well as there is a lot of space between the balls. Also, they are much more expensive when compared to other mediums of the hydroponics system.

Oasis Cubes

They are manufactured from floral foam and are designed in sheet form. Each medium of the individual cells that somewhat looks like small cubes can contain the perfect amount of air, water as well as nutrients for the plant growth. It is being widely used by growers as a beginning environment for plant cuttings or seedlings. So, they cannot be used as a medium of full growth. This medium comes with a neutral level of pH and can easily absorb air and water. That is the reason why it is a great choice for cuttings and seedlings. Also, the plant roots can easily grow and then expand in the open cellular structure of the medium. It is not much expensive as well.

Starter Plugs

This medium is a great choice that is being used widely for starting plant propagation or germination of seeds. You can also find starter plugs that are made out of organic compounds. They can hold the moisture properly and does not get waterlogged. It also permits the plant roots to drive through and expand right from the bottom. But, one downside of this medium is that they are quite expensive and cannot be used as a full growing medium.

Rice Hulls

Rice hulls are the byproduct of rice that is generally thrown away. But, they can be used as a great medium for the hydroponics system. But, they do not come with a neutral level of pH and range between 5.6 to 7.1. You need to remember that fresh rice hulls cannot be used as there might be chances of decaying bugs, microorganisms, and weed seeds. They can drain water very well. But, the downside of this medium is that they get decomposed after a while, and so are needed to be replaced often.

Growstones

They are made up of recycled glass that is collected from glass processing places and landfills. They are crushed, melted, and are often mixed up with calcium carbonates. They are very light in weight and are also highly porous in nature. They can provide great aeration of air and provides moderate retention of moisture to the system of plant roots. You are needed to use them after washing them thoroughly for removing any kind of dust and small particles. But, they are quite expensive when compared to other media.

Sawdust

This is a byproduct of the sawmills. They are quite cheap and are also light in weight. They come with great water retention quality. But, they are not pH neutral and so extra care is needed to be taken before using them.

Chapter 2: Advantages and Disadvantages of Hydroponics

Most people think of hydroponics of being an impossible task. But, hydroponics comes with the power of doing a lot of things. Hydroponics has enabled both the commercial as well as domestic growers to grow plants of their choice in a variety of new ways that come with potential advantages and certain disadvantages. Hydroponics is all about increasing agricultural efficiency along with yield with a decrease in the overall cost of food cultivation. In this chapter, you will be finding the advantages and disadvantages related to this advanced form of growing plants.

Advantages

Hydroponics comes with a wide range of advantages over the traditional form of cultivation. Let's have a look at them.

Better Allocation of Space

All those plants that are grown hydroponically need 20% lesser space when compared to all those plants that are grown in soil. This indicates that you have the power of planting more plants and that too in a very confined space, or you can cultivate plants in a very constrained space where it is not possible to grow plants in soil. This comes along with certain dramatic nature of implications over the industry of farming, where several numbers of plants are being grown in expensive greenhouses, where proper use of space is necessary for achieving a hefty amount of return from the investment.

The main reason behind hydroponic plants taking less space is because the cultivation is carried out without the use of soil, and the plant roots are not needed to spread out much which is in the case of soil cultivation in search of water and nutrients. Nutrients and water are provided directly to the plant roots, either constantly or intermittently, relying on the requirements of the technique. As a result, the roots are more compact in nature and thus can be grown closer to each other. As there is less requirement of space, growers have the power of producing higher yields and that too with less amount of infrastructure.

No Need for Soil

The concept of growing plants without the use of soil was once treated as a foreign idea, but now, it has turned into reality for both the commercial as well as domestic growing. When you grow plants without using soil, there are lots of benefits. Soil quality varies from one place to another, and there are several plants that come with strong compatibility with only one specific type of soil. When you do not have the specific soil type available, the cultivation process might turn out to be extremely expensive for you. Also, people living in the arable type of lands such as rocky islands or deserts will no longer be prevented from growing their own plants. This is the main reason why it is being called as the farming of the future.

Water Saving

Plants that are grown using hydroponics system uses up only about 5-10% of water when compared to plants that are grown in soil. This is an excellent benefit for all those areas where there is water scarcity. This is one major benefit of the environmental aspect that comes along with the use of hydroponics. This whole process relies on recirculated water in which the plants only absorb the quantity of water needed by them, and the run-offs are collected and then returned back to the system. The only form of water loss is via evaporation and water leaks which can be also be minimized by building up an efficient system. Some systems of hydroponics cultivation are using up advanced technologies to reduce the waste of water.

All of the water that is absorbed by the plant roots, only 5% is used by the plant, and the rest is transpired in the air. As a result, some of the commercial systems of hydroponics have started to use condensers of water vapor for recapturing the lost water and then return the same back to the system. There has been a rise in the global production of food every year and is using up more amount of water than earlier days. This is where hydroponics comes into play which can effectively reduce the water loss and use up less amount of water for cultivation.

Climate Control

The hydroponic environments can provide you with complete control over the climate. You have the power of adjusting the intensity of light, temperature, composition, and also air composition according to the need of the growing plants. This ultimately results in the creation of an avenue for producing plants no matter what season it is. Thus, the farmers can now maximize their overall production, and consumers can have access to their required plant products all year round.

Speed of Growth Is Faster

What is the most surprising thing about the hydroponics system is the growing capacity. You might have the notion that hydroponic systems result in lower yields, but in actual, it is the complete opposite. You can grow your plants much faster in a hydroponic system when compared to plant cultivation in soil. This is mainly because you have complete control over the necessary moisture, temperature, nutrients, and light. When you can create an ideal growing condition for the plants, the plants will be taking in the right amount of nutrients that are in direct contact with the roots. The plants will not be using their energy in searching around for water and nutrients. So, all the focus can be given on the nourishment of plants and on the production of veggies and fruits. Thus, the rate of growth is much better and the size of the plants is also large.

Complete Control Over Level of pH

One of the common mistakes made by most of the growers is that they tend to overlook the pH levels. But, the pH level is very important for the proper growth of plants and for ensuring that the plants can access the required quantities of nutrients that they need for growing healthy. Unlike cultivation in soil, the essential nutrients for plant growth are contained in the solution of a hydroponic system. You can easily adjust the pH level of this nutrient solution and make sure that an optimal level is maintained all the time. As you can ensure an optimal level of pH, it will be enhancing the ability of the plants to take in the required minerals. Although there are certain plants that can only thrive in slightly acidic environments, the level of pH needs to range between 5.6 – 7.5. Right before starting with hydroponics, you can study the pH level requirements of the plants for creating the ideal environment.

No More Pests, Diseases, or Weeds

One of the common problems that come along with soil cultivation is weed. They are very hard to remove and can have a great impact on plant growth. With the use of a hydroponics system, there is no risk of weeds, as soil is not used in this system. So, soil-borne pests will no longer be a problem for your cultivation. As the environment is free from soil, the hydroponic systems do not need any kind of application of pesticides and thus, the yields of such cultivation are healthier in nature. Also, there is no risk of soil-borne diseases that might hamper the yield as well as plant growth.

Less Labor

It is true that the overall costing of the hydroponic system is much more than the traditional method of growing plants; no matter whether you use it for commercial or domestic purposes, the overall labor that is involved in this system is much less. This ultimately frees up more time to concentrate on other kinds of activities such as setting up a better system and so on. Also, the cost of running tends to decrease with time.

Weather Is No Longer a Problem

Whether you have adopted a hydroponic system for domestic growth of plants or for larger sized commercial cultivation, you can easily cut off a major kind of uncertainty that comes in the growth of plants. The major uncertainty of plant growth is the weather. As the majority of the systems of hydroponics are indoors, you can eliminate the uncertainty of the weather, provided the required amount of nutrients and water is given properly. Also, the sunlight is no more a concern for plant growth as artificial lighting systems are used in hydroponic systems that can easily replicate sunlight. Using artificial lighting systems can allow you to grow plants of your choice all over the year.

Disadvantages

Just like everything else in this world, hydroponic systems also do come along with certain disadvantages. Although the number of disadvantages is quite less than the advantages, they are major in nature. Also, the disadvantages that you are going to learn in this section might vary according to the system of hydroponic that you use. Let's have a look at some of the major disadvantages of this system.

Needs Your Time and Commitment

Just like other things in your life, a responsible and hard-working attitude can provide you with a satisfactory form of results. In the case of soil cultivation, the plants can be left on their own for few days to several weeks, and they will still be able to survive. Nature and the soil regulate on their own when something essential is not balancing properly. But, that is not at all the case with hydroponics. The plants will tend to die very quickly when proper care is not given and also because of the lack of required knowledge about the system. You are always required to remember that in a hydroponic system, the plants rely on you for their own survival. You need to take good care of them and install an efficient system. When you have an efficient system, you can easily automate the whole thing at a later stage. But, it does look out for a lot of your time along with the proper commitment to the same.

You cannot just leave the system on its own. If you do that, you will end up killing the plants and thus destroying the whole system of cultivation.

Experience Along With Technical Knowledge

In a hydroponic system, several types of equipment work together, at the same time. All that it needs is proper expertise in the whole thing and also the knowledge about how to make plants grow. You are also needed to know how to make the plants survive in case of any emergency situation. Most of the failures that can be recorded in relation to the hydroponic system are because of a lack of knowledge and expertise. But, in the case of soil cultivation, no prior knowledge is necessary. Even a small kid can grow a plant and make it sustain on its own. So, you need to gain adequate knowledge first and learn how the different types of equipment work.

Risks of Electricity and Water

In a system of hydroponics, you are required to use water and electricity in close proximity to each other. So, this actually increases the risk of any major accident as you need to work with electricity and water at the same time. Safety becomes a major concern in this case. Also, in case of a power outage, the system will just stop functioning, and you will end up having dried out plants. So, in addition to an efficient system, you are also required to arrange for power backup that ultimately increases the overall cost.

Initial Costs

You are required to spend a lot of money at the first stage for purchasing the required equipment and for setting up the system. No matter what type of system you are opting for, you will surely need a pump, lights, nutrients, growing media, containers, etc. This makes the whole thing a lot more expensive than the traditional method of cultivation. Yes, it is true that once you have set up the system and it starts running, the cost might get reduced, only electricity and nutrients. But, still, it is an expensive system.

Diseases Might Spread Out Quickly

You are growing plants in a system that is closed in nature and based completely on water. If there is any kind of plant infection or disease, although the chances are less, they might grow very fast as all the plants are being grown in a single reservoir of nutrients. So, the infection or diseases can spread real quick from one plant to another if proper care is not taken. Also, when such a thing happens, it might actually be very tough to actually resolve the problem.

Hydroponics is actually a great hobby. Those of you who love gardening can start by growing some small plants on the windowsill. As you start with a small system, you can eventually expand it. Talking about the disadvantages, they can be outnumbered with a wide range of advantages that is provided by a hydroponic system. Also, when a system is maintained properly, the disadvantages can also be reduced. In fact, the food that is grown using a hydroponic system is much healthy than the food grown using soil cultivation. You can regulate the nutrient proportion by altering the solution that you use.

The most convenient thing about a hydroponic system is that you can have the ability to enhance the nutrients of a plant by adding in all those that are required in the solution. You can start adding whatever nutrients are needed, whether zinc, calcium, magnesium, or iron, to name a few. In this way, you can create a superior quality of plants. It is very much important to note that there are several external factors that you need to take into consideration, no matter what type of system you are using, including the harvesting time, the time period after harvesting, and also the way in which the plants are handled.

Is Hydroponic System Recommended?

Absolutely, hydroponics is a great way of cultivation. Yes, it is true that it comes along with certain downsides like everything else in this world. But, a majority of them can be overcome most of the time with some proper planning and knowledge. Taking into account the advantages of this system, hydroponics is worth trying your hands on. The potential of the same is also increasing day by day. The overall market concerning hydroponics is expected to rise drastically in the upcoming years. If you really love gardening and cultivation, it is worth the expenditure.

The overall business of cultivation is most likely to shift to this system owing to its high quality of yield and greater production. In fact, many commercial growers have already started using this system. The rate at which the market of hydroponics is increasing, that day is not far when this system will be taken as a viable cultivation technique. Hydroponics will also allow you to try your hands on various types of plants as it can support the wide variety of plants with ease. Just give in some of your time and concentration and you will be able to create a really great system for cultivation of your own plants. Also, there are various types of hydroponics that will be discussed in the next chapter.

Chapter 3: Types of Hydroponics

When it comes to the hydroponics system, it is of several types. In this chapter, we will be discussing some of the most common types of hydroponic systems that are being used today.

Deep Water Culture Systems

In deep water culture hydroponic, the plants are simply suspended in aerated water. Also known as DWC, it is often regarded as one of the easiest and also the most popular type of hydroponic system that is available in the market today. In such systems, the net pot plant holding dangles over a solution reservoir containing oxygen and nutrients. The roots of the plants remain submerged in the nutrient solution while providing the same with water, oxygen, and nutrition. This system of hydroponics is regarded as the purest type of hydroponics. As the system of roots is suspended in the solution all the time, the proper form of oxygenation is very much important for the survival of the plants.

In case there is not enough oxygen supply to the roots of the plants, there are high chances that the plants will be drowning in the solution. For proper oxygenation of the whole system, it is best to add air stones along with an air pump right at the base of the reservoir. The air stone bubbles will also help a lot in circulating the solution of nutrients.

It is not much tough to assemble a system of this type at your home as it does not any kind of expensive equipment for the same. You just need a clean bucket or an old aquarium for holding the solution or use the same as a reservoir. You need to place a floating type surface on the top of the solution that will be holding the net pots. The plants in this system can only have their roots submerged in the solution. No other part of the plant can be underwater. Try to leave approx one inch or half of the plant roots above the surface of the water.

Advantages

- After setting up, it needs very less maintenance. All that you need to do is to replenish the solution of nutrients when necessary and just ensure that the pump is providing the required amount of oxygen. You will typically need to replenish the solution every two to three weeks.

- Unlike any other system of hydroponics, DWC can be created very cheaply and is also easy to set up. You just need some very basic equipment, and you are good to go.

Disadvantages

- This system is very well adapted with growing lettuce and small herbs. It cannot be used for large-sized plants. It is not at all meant for something that flowers.

- You need to maintain the temperature of the water all the time, not more than 67 degrees Fahrenheit and not below 60 degrees Fahrenheit. As the water in a static state, it might be a tough job to properly regulate the temperature.

Wick System

In this system, the plants are placed in media of growth and are placed on trays that sit right at the top of the water reservoir. The reservoir holds water that comes with dissolved nutrients. Wicks tend to travel right from the water reservoir to the tray. Nutrients and water travel up the wick and get saturated in the growing media near the system of roots of the plants. Wicks can be made out of very simple things such as string, rope, or even felt. It is also a simple system of hydroponics. The wick system is actually passive hydroponics. In simple terms, this system does not need any kind of mechanical working parts, such as pumps for functioning. This makes this system an ideal one for all those situations in which there is no availability of electricity or is unreliable.

This system works through the process of capillary action. The wick tends to absorb the solution in which it is immersed, just like a piece of sponge. When it comes in close proximity to the growing media, the nutrient solution is transferred to the media. But, this system of hydroponics will only work when it is coupled with a medium of growth that can actually facilitate water and nutrient transfer. Coconut coir comes with superb retention of water and also comes with the extra benefit of having a neutral level of pH. There are other growing pH neutral growing media as well that can be used as a medium in the wick system.

But, you need to note that the wick system is the slowest of all other systems of hydroponics. This, in turn, limits all those things that can be cultivated with this system. You need to ensure that for each plant in the tray, there is a minimum of one wick that runs from the nutrient reservoir. Also, the wicks are needed to be placed in close proximity to the system of roots. Although this system can function with the help of aeration, many growers all add up an air pump with an air stone for adding in extra oxygenation to the overall system.

Advantages

- This system can be set up by any individual and does not need any form of extra attention after setting up. There is no risk of the plants drying out as the wick system will be constantly supplying water to the plants.

- This system can be installed at any place and does not need any extra space for its running.

-

Disadvantages

- Fast-growing plants such as rosemary, lettuce, basil, and mint do not need a large amount of water and also grows fast. Other plants, such as tomatoes, might really struggle in thriving in such a system as they need a large number of nutrients along with hydration. Also, plants such as turnips and carrots cannot thrive in this system of hydroponics.

- Wick systems always remain damp and humid. This might end up in the outbreak of fungal action and might rot the plant roots.

Nutrient Film Technique

Also known as NFT, suspends the plants over a continuous flowing stream of nutrient solution that touches and washes along the end part of the root system of the plants. The channels in which the plants are placed in a tilted manner so that the water can easily run down along the overall length of the tray right before draining into the nutrient reservoir situated below. The reservoir water is kept aerated with the help of air stones. This system uses a submersible pump that can effectively pump the water right out of the reservoir and then back to the channel top. This is a recirculating system of hydroponics. Unlike the deep water culture, the plant roots are not submerged in the nutrient solution. Instead of that, a stream of water flows over the root ends.

The tip of the roots will be taking up the moisture into the plant. The exposed part of the root is provided with an adequate supply of oxygen. The base of the channels comes with grooves so that the water film can easily travel across the tips of the roots. This also helps in preventing the water from damming or pooling the root systems. Although in this system, the water is being recycled constantly, it would be great if you could drain out the reservoir and then replenish the solution of nutrients after every one week or so. This will be ensuring that the plants are getting an adequate amount of nutrition. The channels of the NFT system need to be angled at a slope that is gradual. In case the angle is too steep, the water solution might tend to rush down the provided channel without proper nourishment of the plant roots. When excessive water is pumped across the channel, there are high chances of the system getting overflowed and ending in the plants getting drowned.

This system of hydroponics is very popular among commercial growers. You can easily grow a large number of plants in each channel and thus allowing mass production. This system works the best for all those plants that are light in weight, such as lettuce, spinach, kale along with fruits like strawberries.

Advantages

- As the water is recirculated in this system, there is no form of high demand for huge quantities of nutrients or water for functioning. Also, as there is a constant flow of water, salts cannot accumulate at the base of the plant roots. This system does not need any form of growing media, and thus, you can save a lot in this aspect.

- This is a great choice for large-scale growers. After setting up the system, you can expand the same very easily. You can easily fill up your greenhouse with several channels containing various types of crops at a time. It is always suggested to provide one reservoir each for each channel.

-

Disadvantages

- In case the water pump fails and the channel is not able to circulate the film of nutrients, the plants will tend to dry out. Only within a few hours, the plants can die if the proper supply of nutrients is not provided.

- When the plants are placed in close proximity to each other, there are high chances of the channel getting clogged. If there is a clog in the channel, water will not flow properly and will result in starving plants.

Ebb and Flow Hydroponic System

This system works by flooding a growing bed with a nutrient-rich solution from the reservoir that is placed below. The submersible pump that is being used with this system comes along with a timer. As the timer starts, the grow bed gets filled with nutrients and water supplied by the pump. As the timer sets off, the water is drained out slowly from the grow bed and is flushed back again in the reservoir. This system also comes along with an overflow tube that can make sure flooding does not exceed a certain level, and no form of damage is done to the plants and fruits. Unlike all other systems that have been mentioned above, the plants that are used in this system are not exposed to water constantly. As the grow bed gets flooded with the nutrient solution, the plants take up the required nutrients via their system of roots.

When the water is drained back to the reservoir, the roots of the plants dry out. The dry plant roots then try to oxygenate in the break right before the upcoming flood. The total time that is kept between the floods is decided by the grow bed size and also by the plant size. This is one of the most popular forms of hydroponic systems. As the plants are supplied with abundant nutrition and oxygen, the growth of the plants is quick and vigorous in nature. You can also customize this system very easily and is versatile in nature. You can fill up the growing beds with net pots and also with a great variety of veggies and fruits. The best part about this system is that you will be able to experiment with your setup.

The ebb and flow system is able to accommodate any type of plants. The primary form of limitation that you are going to face is the depth and size of the trays. Some of the popular crops of this system are beans, peas, tomatoes, carrots, and cucumbers.

Advantages

- With the help of this system, you will be able to grow large-sized plants. Veggies, fruits, and flowers, all respond very well to the system of ebb and flow. If have set up a proper system for the plants, you will be able to see a great yield in the coming days.

- You can create your setup in various ways. You can design the system according to your needs.

Disadvantages

- Just like any other system of hydroponics that uses up water pumps, this system will also fail to work when there is a pump failure. You are needed to constantly monitor the working of the water pumps to avoid any kind of problem. Also, if the water tends to flow out very fast, the plants will not be receiving the proper amount of nutrition as needed.

- Two of the primary requirements of this system are maintenance and sanitation. If you fail to do so, the root might catch diseases with setting in of rotting. Also, when the system is dirty, it can easily attract pests and other insects. When cleanliness is neglected in this system, the plants might end up suffering.

Drip Hydroponic System

In a drip hydroponic system, the nutrient-rich and aerated reservoir pumps the solution via a network of tubes that are dedicated to each plant. The nutrient solution is dripped very slowly in the medium of growth that surrounds the system of roots. This helps in keeping the plants well-nourished and moist. This is a very popular system related to hydroponics, specifically among all the commercial nature of growers. You can use a drip system either for an individual plant or for a large operation of irrigation. In a hydroponic drip system, there are two types of configurations that can be found: recovery and non-recovery. In the systems of recovery, which is most commonly used by the home growers, the excess amount of water gets drained right from the base of the growing bed right back into the solution reservoir. The same solution is again recirculated for the upcoming cycle of dripping.

In the system of non-recovery, the excess amount of water is effectively drained out from the base of the growing bed and is thrown away as waste. This configuration is more common among the commercial nature of growers. Although this configuration of the drip system might seem to be wasteful, the large-scale cultivators are very much conservative about the usage of water. This type of drip system is used for precisely delivering the required amount of solution for keeping the medium of growth around the plant's root system properly moistened. This system does not come along with any type of timer or schedule of feeding for keeping the waste size minimum.

If you are opting for a recovery type of drip system, you are required to be in pace with the fluctuations in the level of pH of the solution. This actually applies to any kind of system where the waster form of water is again recirculated in the system. The plants will tend to deplete the content of nutrients and might also alter the balance of pH. So, you need to properly monitor the system and adjust the content of the reservoir solution. Also, there are high chances of the growing media getting oversaturated with the supplied nutrients. So, you might also need to wash them and replace them from time to time.

Advantages

- While using a drip system, you have the advantage of supporting large-sized plants that cannot be supported by most other systems of hydroponics. That is why this system is so much popular among commercial growers. Also, drip systems can hold up more quantities of media for growth than any other system. Thus, it can easily support larger systems of roots.

- If you want to add up more plants to your existing system, you can do that very easily. You just need to add up new tubes to the reservoir, and you are good to go. You can also add in additional reservoirs if needed for large-sized growing.

-

Disadvantages

- When using this system for growing plants, you need to maintain the system properly, especially in the case of non-recovery systems.

- This is a complex system of hydroponics and requires high expertise for the effective growth of plants.

-

Aeroponics

In this system, the plants are suspended in the air, and the naked root systems of the plants are exposed to a mist that is filled with nutrients. T

his system works in enclosed frameworks such as towers or cubes. This system can easily hold up many plants at one time. The nutrients and water are stored in a large reservoir just like other systems. The solution is pumped to a mist nozzle that atomizes the nutrient solution and then distributes the same as a fine mist. This system uses up less water when compared with other systems of hydroponics.

Advantages

- This system can be customized according to your own needs. Also, the surplus amount of oxygen that is taken by the system of roots helps in supercharging the growth of plants.

- It uses up less water than any other system.

-

Disadvantages

- They are quite expensive than other systems. You will need various types of equipment, such as timers, pumps, reservoirs, etc.

- It is very tough to maintain this system. If the functioning of the system gets disrupted by chance, the results are going to be disastrous for all the plants in the system.

Chapter 4: Features of a Hydroponic Garden

Most of the people today are very much interested as well as excited about the system of hydroponics for growing plants. But, most of them do not even know about the features of the same. Right before setting up your system, it is very much important that you learn about all the features and equipment first and then start with the process. So, in this chapter, you will be learning about the primary features that you need to consider before setting up a hydroponic system.

Water Pumps

Water can be regarded as the soul of a hydroponics system, and without the presence of water, plants will not be able to survive. A water pump is used up for circulating all the water to the various parts of the hydroponics system where the plants are situated or located. Yes, there are certain ways in which a water pump is not needed because of the system of working. Such systems include wicking system and water culture system. In both of the systems, there is a stationary reservoir of water from where the roots of the plants take in the required nutrients directly from the given water. We will learn about these in the next chapter.

When you are using a system in which a water pump is necessary, there are certain differences that can be used to meet all your requirements.

- **Submersible:** This type of pump sits inside the reservoir of nutrients and remains covered by water fully. This is the most common type of pump that is being used for the systems of hydroponics. This type of water pump is much cheaper when compared to others and are also easy to install. The best thing is that this pump is quiet and does not make any kind of sound. But, the downside of this pump is that it generates heat while working, and sometimes this can result in rising up the temperature of the nutrient solution.

    Also, this type of pump is not at all meant for large systems where the need for water is 1000 gallons every hour.

- **Inline:** This type of pump is different from the previous one, and it sits outside the water reservoir. Inline pumps are generally used for larger systems. This type of pump is not that easy to use unlike a submersible pump and also makes a lot of noise. Inline pumps are also expensive.

But, these pumps do not generate much heat that can actually pass into the nutrient reservoir. Also, these pumps can pump a larger quantity of water.

While choosing water pumps for your system, there are various things that you are needed to check. Such things are:

- **Head height**: This is the total distance right from the base of the reservoir to the top of the water that you want to reach. It might actually be a difficult task to find out the measurements, and so it is always advisable that you select a large size pump than you actually need. This will also provide you with the capability of expanding the system with no more need for any extra pump. Also, you are required to check the efficiency of the pump that you choose as not all pumps are equally efficient like the others.

-

Air Stones and Air Pumps

Right after water, one of the most crucial components that are needed by plants is oxygen. If the water in the reservoir tends to become stagnant, there will be no oxygen, and the plants might drown. With the use of air stones and air pumps, you can easily diffuse air into the reservoir water. Also, with the development of bubbles by the air pumps, the solution can be mixed continuously. Just like the water pumps, it is also very tough to find the proper size of the air pump. As a standard, it is always better to have an air pump that can deliver 600-700 cc every minute. As air pumps generally sit outside the reservoir, they can make a humming sound continuously that might become annoying at times. For getting the quietest one, you can check the decibel levels of the pumps.

No matter if you have a small system of hydroponics or a large one, it is always suggested to use an air pump that comes along with several nozzles. This will not only help in spreading the air all across the reservoir, but when you have several growing units, you can very easily place air stones inside the other tanks without any need for splitting the tubes. Another thing that is worth noting is the tubing that you are going to use. You might get clear tubes, but it might not be a great idea as light can easily seep into the reservoir and might result in the growth of algae with the building up of moisture.

While choosing the air stones, it is always better to choose the ones that can produce small size bubbles. With small bubbles, it can easily get exposed to a larger surface area of water. Also, when the size of the bubbles is small in size, they will tend to travel much slower, so resulting in much more oxygenated water.

Reservoirs, Flood Tables and Trays

Reservoirs are very much important in a hydroponics system, and based on the system. They can be used in various ways. When you buy growing systems, they generally come with reservoirs, but they will not permit you to expand as the reservoir is matched only with the system. A suitable size of the reservoir depends on several factors that you need to consider. Humidity is one of the variables that get overlooked most of the time. The normal level of humidity that is suitable for plant growth is around 70 – 80% and if that falls below 40%, it is the indication that the plants will need to take in more nutrients and water.

You can use a very simple formula for determining the reservoir size for the system. Here are some of the requirements for each size of the plant:

- Small plant: minimum ½ gallon

- Medium plant: minimum 1 – 1 ½ gallon

- Large plant: minimum 2 ½ gallons

While considering a reservoir, make sure you pay your attention to the lid of the same. It is very important for limiting the amount of evaporation and can also help in preventing the growth of algae. Lids also help in limiting any kind of light from entering the mix of nutrients. Another thing is the temperature of the water. Always try to maintain a temperature of 70 – 75 degrees. A large-sized reservoir will be able to maintain its temperature much easier than a small reservoir that might result in fluctuating the mix of nutrients. You are also required to clean the reservoir daily for preventing any form of growth and for preventing the building up of bacteria.

Grow Trays and Flood Tables

When it comes to grow trays and flood tables, they are more or less similar in construction and are made from a very durable nature of plastic. Some of them come with pre-drilled holes for allowing draining back of water to the reservoir. The differences are in the drain and flood systems in which the sides are tall enough to accommodate a considerable amount of water. When flooded, they soak the growth medium right before draining when the cycle of timing ends. While using these for building up drain and flood systems, they are very cheap to make and also need a very small quantity of equipment to build and run.

The trays that are shallower in depth are ideal options for the seedlings along with the small-sized plants.

Lighting System

Right after oxygen and water, the next thing that comes in the list of importance is light and is a very important thing for the survival of plants. For the systems of indoor plant growing, you can have several options for lighting. Also, the lighting might turn out to be one of the toughest things to control as each light type delivers several benefits while also coming along with several downsides. One thing that you need to understand is that any system of lighting will have to replicate the same thing that is being delivered by the sun all throughout various seasons, and also for different stages of growing plants.

HID

Also known as high-intensity discharge, it falls under two primary sections and can be either HPS (high pressure sodium) or MH (metal halide).

MH (Metal Halide Bulbs)

These bulbs are generally used during the phase of growth as they come with the capability of producing wavelengths that are either towards the white or blue end of the spectrum.

Thus, it can very easily simulate the sun of the hot summer. Some types of plants, like leafy vegetables and herbs, can reach their maturity with the help of this bulb only. There are several plants that can also flower under this type of light but the overall yield will be very low. So, this type of light is generally used in conjunction with the HPS type of lights.

HPS (High Pressure Sodium Bulbs)

The wavelength of light produced by this type of bulb is in the red end of the spectrum. So, this type of light comes along with a tint of reddish-orange, which can easily simulate the warmer colors right during the period of harvesting. These lights are generally used when the plants start flowering. This combination of light is actually very effective in nature as it can very easily replicate a complete growing season for most types of plants. This type of light is also very cheap but comes along with one disadvantage. These lights tend to produce extreme heat and also use up a lot of electricity for running.

Fluorescent Lights

These bulbs can emit cool light and is perfect for seedlings and cuttings. It produces a very low amount of heat while delivering light at a wide-angle. The smaller version of this,

CFL, comes along with a combination of more than one bulb. The ballast comes along with this system of lighting and thus makes the system perfect for all those growers who have smaller systems. Also, this type of light is also very cost-effective in nature. While these lights might not have the power of lighting when the plants are in the abundant stage, these lights can only be used for the growth of small seedlings.

LED

This system of lighting is gaining its popularity gradually. Although this system costs much more than any other system of lighting, they are actually very much energy efficient and there is no need to change the bulbs for a couple of years as they last very long. They are very cool when it comes to operation and they are very much perfect for the small growing systems as they will hardly be affecting the surrounding temperature of the system. They can be placed in close proximity to the plants as there is no worry of burning them or drying them out. LEDs that can produce spectrum can be bought or you can also opt for the multi-colored ones that can easily cover all types of the spectrum as needed for the growth of plants. Thus, this system makes the whole thing very straightforward.

Sulfur Plasma

Just like the new systems of LED lights, these are also completely new to the system of hydroponics. This system can come along with various ranges of outputs that can be adjusted very easily. The range that is supported by this system is from 100W right up to 1300W. Thus, the growers can use this system of lighting for multiple types of plants and that too in various sizes of rooms. Right at the moment, the overall effect of this lighting system on the systems of hydroponics is not known fully. But, it is being claimed that this system comes with superb energy efficiency and it can easily replicate the natural form of light to a great degree of accuracy as other systems of lighting.

Chapter 5: How to Set Up a Hydroponics System

Most of the hobbyists tend to opt for the system of hydroponics as they just want to grow their own plants and foods and also because they do not have much access to any outer space. Although the sun is regarded as the ideal source of lighting for the proper growth of plants, an artificial system of lighting in a hydroponic system can actually provide you with a great substitute and that too with the proper spectrum of colors.

How to Choose the Perfect Light?

As a beginner, finding the perfect light for your system might turn out to be a daunting task. There are various options available when it comes to lighting that has already been discussed in the previous chapters. But, the perfect light would depend on the size of your system and also the plant type that you intend to grow in the system. In the traditional way of cultivation outside, a garden of veggies needs about five to six hours of direct light from the sun every day along with at least 10 hours of indirect light. In a hydroponic system, the primary goal is to imitate this type of lighting condition. You need to plan the system for having a minimum of sixteen hours of artificial light that is bright in nature along with 11 – 12 hours of darkness per day. Remember, that darkness is equally important as light.

In case the plants that you are growing are perennials, you are required to be extra careful about the schedule of lighting that you provide to the system at the stages of growing and flowering. The best way of maintaining a proper schedule of lighting is by using an electronic timer. They are actually worth your expenses as if you ever forget to turn on or off the lights; it might result in some adverse effects for the plants.

Different Plants and Different Needs of Light

If you are growing various types of plants in a system, an electronic timer can save you from a lot of hassle. You can set a timer according to the needs of the plants or as the garden grows.

Short day Plants: Some plants need extended periods of darkness for photosynthesizing and for producing flowers. In case such plants are exposed to light for more than 12 hours, flowering might not take place. Some examples of short-day plants are strawberries, poinsettias, cauliflower, etc. The cycle of short day mimics the natural environment for the plants for flowering during the time of spring.

Long day plants: Some plants need a long time of sunlight, around 18 hours, every day. Such plants include potatoes, lettuce, turnips, spinach, etc. The long day cycle helps in mimicking the preferred natural conditions for the plants that flower during the summertime.

Day-neutral plants: Such plants are the most flexible ones. They can produce flowers or fruits, no matter how long they remain under the light. Such plants include roses, eggplant, corn, etc.

Parts of Lighting System Necessary for Hydroponic Setup

All major systems of lighting that are used for hydroponic systems come with four primary parts.

Bulb: The most recommended wattage for any bulb that is used in a hydroponics system ranges between 500 – 600 watts. Most of the growers try to opt for HID lights. The bulbs of HID can produce light by passing an electric arc between two pieces of electrodes that come encased in a structure of glass along with a mixture of metal salts and gas. In an HID lighting system, there are two types of bulbs: HPS and MH. The MH bulbs can act as great all-day light. If you have room for only one bulb, this is a great choice.

The HPS bulbs are used mostly during the stages of fruiting or flowering. They are expensive than the MH bulbs and most often used in partnership with the MH bulbs.

Reflector Hood: It is nothing but a type of reflective casing that is put around the bulb of the lighting system. It helps in increasing the efficiency and effectiveness of the lighting system by reflecting the light on the plant surfaces from various angles. It helps in providing an effective light spread. Also, by using this, you can easily cut off the excess heat that is produced by the bulbs of the lighting system. It can also help you in saving a lot of electricity by increasing the spread.

Remote Ballast: It is the power box that helps in providing the required power to the lights. Sometimes they are sold along with the lighting system assembly, but they are quite heavy in weight and give out a lot of heat. They are recommended for the home-based systems. This is also the most expensive part of the entire lighting system, and so they are needed to be kept away from the solution so that they are not damaged in case of any leaks or flood. If you are using the ebb and flow system, flooding might turn out to be a common scenario. It is suggested to get the ballast along with the lighting system as they need to match the wattage of the system as well.

Timer: They are the cheapest part of the lighting system but also the most essential one. They need to be of heavy-duty and are also needed to be grounded. You can choose either the electric one or the manual one. The manual version comes along with pins and two plugs on both the sides for attaching to the lighting system at once. The manual ones are more often used than the electric ones as they come with less chance of breaking. It can help you in setting the schedule of lighting according to the need of the plants.

Nutrients as the Core of a Hydroponic System

Hydroponic systems can help a lot in conserving water as they need very less amount of water and are also water-based. Water is used as the main method of delivery to the plants. The required nutrients are mixed up in the water and are directly available for use to the plants. But, proper management of nutrients is necessary for the optimal condition of plant growth.

Great nutrient management of a hydroponic system takes place when the growers are:

- Completely aware of the plant nutrients and their point of origin.

- Providing the required quantity of nutrients to the plants.

- Supplying nutrients in the proper ratio to the plants.

- Measuring and monitoring plant nutrients from time to time.

- Making workflow-oriented decisions regarding nutrients.

-

16 Plant Nutrients

Most of the plants rely on 16 different types of nutrients that help them in growing and reproducing. Out of these 16, three of them are available via uptake of water and also via gas exchange: hydrogen, oxygen, and carbon from CO_2. Growers tend to think about the movement of air and also levels of oxygen in water, the timing of irrigation, etc. , but all of these mentioned activities are most of the time given importance separately from the management of hydroponic nutrients.

The remaining 13 nutrients are the mineral-based nutrients that are delivered to the concerned plants by dissolving the same as a nutrient solution. All of them can be divided into three separate groups:

- **Primary macronutrients:** They are the most widely found building blocks that are responsible for the growth of plants and reproduction of the same.

- **Secondary macronutrients:** They are required in small amounts and are equally important.

- **Micronutrients:** They are needed in a very small amount for reproduction and growth.

The Primary Macronutrients

The primary macronutrients that are needed for plant growth are phosphorus, nitrogen, and potassium, popularly known as NPK.

Nitrogen is necessary for all forms of molecules that are related to photosynthesis and also in the creation of protein. It can be supplied at once either as a liquid fertilizer or by dividing it into two different parts: mix of NPK and $CaNO_3$ as dry fertilizers. Phosphorus is essential for the cell membranes and can be supplied through the main mix of nutrients, either liquid or dry. Potassium is the primary key for the signaling compounds that are used for the growth of plants and the development of the same in various stages. It is supplied via the main mix of nutrients.

Secondary Plant Nutrients

The secondary nutrients are magnesium, calcium, and sulfur. Calcium is essential for the building up of cell walls and also functions as a beneficial element of the structure. Calcium can interact in a unique way with other important nutrients, is less soluble when compared to other types of nutrients, and can result in precipitation. So, calcium is needed to be mixed up separately. It can be supplied as $CaNO_3$ or calcium nitrate. Magnesium is beneficial for the photosynthetic complex. It is generally supplied as $MgSO_4$ or magnesium sulfate, popularly known as Epsom salt and also with the main mixture of nutrients. Sulfur plays an important role in the peptide bonds and is present in all forms of biological molecules. It is delivered as $MgSO_4$ along with magnesium.

Micronutrients

The important micronutrients are chlorine, boron, iron, copper, zinc, molybdenum, and manganese. Without the presence of any of these micronutrients, plants cannot survive.

Measuring Nutrients Using EC

The complete level of nutrients that is present in a solution is generally measured in electrical conductivity or EC. EC helps in measuring how well a solution can transmit electricity. This actually works out because all the mineral types of nutrients are salts by nature and can dissolve very easily for turning into ions in any solution and the ions tend to make them more conductive in nature. So, it can be said that at times when we measure the solution conductivity, we actually measure the available nutrients in that very solution. A meter is used for finding out the EC that comes with two metal probes for measuring conductivity. A stream of current is passed from one probe to the other in the water solution and then the current strength is measured. The strength is then converted in the form of measurement that depicts the number of salts in that solution.

The units that are used for measuring EC are mS/cm or ppm. ppm is the most widely used unit for measuring the total number of dissolved forms of solids. But, as a grower, you are also needed to learn about the 0ther unit, mS/cm. This is most of the times just denoted as EC level, for instance, the EC of a certain solution is 2.0, with no form of unit.

The ideal values of mS/cm lie between 1.3 and 3.4. There is a wide range of EC levels that are acceptable, and every plant comes with its own ideal range.

Nutrient Formulas and Ratios

All fertilizers are mixed up using certain ratios. Different crops come along with different needs of nutrients at a particular ratio. When you can use the perfect ratio, you can easily avoid any kind of toxicity or deficiency of nutrients. For instance, let's have a look at the lettuce growing nutrient formula:

Total Nitrogen – 8%

Ammoniacal Nitrogen – 0.50%

Nitrate Nitrogen – 8%

Soluble Potash – 35%

Trace Elements:

Boron – 0.30%

Iron – 0.50%

Copper – 0.01%

Manganese – 0.30%

Soluble Manganese – 0.20%

Zinc – 0.04%

Molybdenum – 0.02%

Chlorine – 3.00%

Mixing Up Solutions

The perfect way of mixing solution is by following the instructions of the manufacturer. The manufacturers will always be providing instructions for the proper mixing of the solution. But, if you want, you can also start mixing them up according to your own knowledge. But, make sure that you learn about the nutrient requirement first and then opt for mixing it up according to your choice. When doing it on your own, it is better to use a lesser amount of nutrients than required so that you can easily tackle the mixture by increasing some components afterward.

Setting Up Hydroponic Garden

The very first step that comes in the process of setting up a hydroponic garden is determining the location. You can set up the system either in an enclosed place, such as a greenhouse, in the basement of the house, or even in any outdoor deck or patio. The floor of the location needs to be properly leveled for making sure of even coverage of the nutrients and water to the growing plants in the hydroponic system. If you have decided to place the system outdoors, make sure that you are protecting the system from various elements, like providing a barrier for string wind, checking the level of water at regular intervals for making sure that there is no form of water loss because of evaporation, etc. When the outside temperature falls considerably, especially during the winter months, try to bring the system indoors. If you have decided to set up the system inside your house, such as in any interior room of the house, make sure that you add growing lights for providing the plants with supplemental light.

Step 1: Assembling the System

The most basic and easy type of set up can be made by using six tubes for growing the plants, that is made from approx 6 inches of PVC pipe. You need to make a stand for the PVC pipe growing tubes along with a trellis, a nutrient tank that can hold approximately 50 gallons of nutrient solution, a manifold, and a pump. You need to set the tank right below the PVC growing tubes. Place the pump inside the base of the tank for pushing up the nutrients to the growing plants through a manifold that is made of smaller pipes of PVC and plastic made tubes. Each of the growing tubes needs to have a drainpipe that will be leading its way back to the nutrient tank. The manifold needs to be placed right above the pipes and is responsible for sending in pressurized water through the tubes.

For getting the nutrients right to the plants in the system, water needs to be pushed via a PVC pipe square and then the manifold. The nutrient solution will be shot out to the small-sized plastic tubes that run through the inside portion of the bigger growing tubes. The tube that carries the nutrient solution comes along with tiny holes in its surface, and one hole is placed right between the sites of each plant. The nutrient solution is pushed out of the tube holes and is sprayed to the roots of the plants. At the same point of time, the water jet keeps on making bubbles of air so that the growing plants can get enough amount of oxygen.

Step 2: Mixing Water and Nutrients In the Tank

Start by filling the tank of 50 gallons with water. Then, add in two cups of the nutrients to the water tank or as instructed by the manufacturer instructions. After that, turn on the water pump and allow the system to run for about half an hour for allowing all the nutrients to get mixed up properly.

Step 3: Adding In Plants to the Grow Tubes

One of the best ways of planting a hydroponic based garden is by using the purchased seedlings, especially when you do not have the required time for growing the seeds by yourself. The primary key is to select the healthiest plants that you can get and then follow the same by removing all the soil from the roots of the plants. For washing the roots of the plants, submerge the plant roots in a container of normal temperature water. If you use water that is too warm or too cold, it can send the small plant into a state of shock that can actually damage the plant. Start to separate the root system gently for getting out all the soil. Any tiny particle of soil that is left back in the root system can easily clog up the small holes of spray that are present in the tubes of nutrient solution.

After cleaning the roots, try to pull out as many pieces of the root system through the base of the planting cups. Then, add in expanded pebbled of clay for holding the plants upright in its place. The expanded pebbles of clay are hard and light as well, so they won't be hurting or damaging the roots of the plants.

Step 4: Tying the Plants Down From the Trellis

You can use plant clips along with strings for tying the growing plants with the trellis. The string will help in providing the plants with proper support so that they can climb up straight, which ultimately helps in maximizing the total space in a very constrained area. Tie the plant strings loosely from the trellis and then attach the plant clips along with the strings right to the plant base. Wind up the plant tips gently right around the tying string.

Step 5: Start the Pump and Regularly Monitor the Entire System

After you are done with setting up the system, now it is time to turn on the pump and check the levels of water daily. In some places, you might even need to check the level of water two times a day, relying on the loss of water because of the excessive amount of evaporation and heat. You are also needed to check the level of pH along with the nutrient level after a certain period of time. In this system, as you are using a pump that runs full time, there is no need for a timer, but you also need to make sure that the nutrient tank does not dry out as it might result in burning up the pump coil.

Step 6: Monitoring Growth of Plants

Right after few weeks of planting, the plants will tend to cover up the trellis completely as they will be getting all the nutrients and water that they require for growing fast. It is very important to properly note the growth of plants and try to clip or tie the stalks of plants every three to four days.

Step 7: Inspecting the System for Diseases and Pests

You need to look out for the signs of diseases and pests, such as sightings of insect type pests, chewed up leaves, or discolored leaves. If you fail to notice a diseased plant, it might end up infecting all other plants in the system as diseases and pests tend to spread very quickly in a hydroponic system as a common solution of the nutrient is used for the entire system. Try to remove all those plants that are sick or infected. Although the chances of having pests or diseases in a hydroponics system are very less, there might be some form of infestation when proper care is not taken. When proper care is taken, you can protect the system from all types of diseases and pests.

Even though hydroponic systems are great in fighting off diseases, there might still be the need to fight off pests. Try to treat the plant that you think or find out is infested by any pest. You can use mild or organic insecticides as well. A detailed process of fighting with pests will be discussed in the next chapters.

Chapter 6: Plants Suitable for the Hydroponic Garden

As we have progressed a bit in the aspect of hydroponics, you might have a very common question in your mind 'What plants are suitable for a hydroponic system?' Well, for answering your question, there are various plants that you can grow hydroponically that include both veggies and fruits. Talking about exceptions, mushrooms are one of them as it is not a plant and is a fungus by nature and comes with a different growing process. Also, there are certain plants that cannot even grow when kept in a water-based environment. So, let's have a look at all those plants that you can grow hydroponically, without any kind of issue.

Lettuce

Lettuce is often known as the most popular plant for hydroponics systems, mainly for all those who have just started with hydroponics. The main reason behind lettuce being such a popular hydroponics plant is that it can be grown very easily both indoors and in greenhouses in an environment that is completely controlled. Also, lettuce can be grown all throughout the year, so that is a plus point. For the very first lead harvest of edible form, lettuce takes about three weeks, and for getting a harvest of full-head, it takes only about 50-85 days. So, it can be said that the harvesting time of lettuce is quite short. Also, you can grow lettuce in lighting conditions that are not much powerful as lettuces do not need much amount of sun for its growth.

The solution of nutrients that are needed for lettuce growth is not that complex as it is a leafy vegetable. So, there is no stage of flowering that is involved in the process. There is no need to change the solution of nutrients all over the time of the growing season. Lettuce that comes with loose heads is the best-suited one for growing hydroponically. Some of the most commonly found and hydroponically grown varieties of lettuce are loose-leaf lettuce, romaine lettuce, and butterhead lettuce. Lettuce does not need any kind of extra care and thus, it is a perfect crop for the beginners.

Tomato

Tomatoes are one of the easiest growing crops hydroponically right after lettuce. Having tomatoes in your garden all the time is actually a great thing as tomatoes are super healthy and also contain an important nutrient that is beneficial for the health of your cardiovascular system, known as lycopene. When it comes to tomatoes, there are various varieties that are available, and the majority of them can be hydroponically grown. Two of the most common tomatoes that are grown by the process of hydroponics are traditional tomatoes and cherry tomatoes. Tomatoes are grown extensively by both hobbyists and commercial growers.

Hydroponically grown tomatoes come with a superior kind of flavor and are also filled up with nutrients. Tomatoes are grown by using two common patterns, and it depends on the variety of tomato that you are cultivating. In the first place, there are the bush varieties that are much difficult to maintain and they can even sprawl across the entire floor area of your greenhouse. They are very common in heirlooms. Right because of this, moving around in the greenhouse or even picking up fruits might turn out to be really difficult. Second, are the ones that are of the vining variety. This variety is most commonly preferred by the growers. They can be pruned much easily, and thus, it makes harvesting a lot faster.

Kale

Kale is a very nutritious kind of vegetable and is used widely in various types of foods. Growing of kale in the hydroponics system is easy as well. Kale cultivation in hydroponics systems does not need any kind of extra care, and it can thrive on its own in any system that is based on water. The health benefits related to kale are endless and so are the culinary attributes. But, in the majority of the cases, it is very hard to find fresh kale in the market or in the stores. That is the reason why growing kale in your hydroponic system is a great idea. It can provide you with fresh kale all the time. For getting fully matured kale, it will take around 3 – 4 months in a hydroponics system. If you want to harvest young baby kale, it can be done within one month only.

Beans

Beans are of various types, and most of them can be grown in a hydroponics system. They come with zero requirements of maintenance and are thus very easy to grow. Some of the varieties of beans that can be grown easily in a hydroponics system are bush beans, runner beans, lima beans, and string beans. The time required for germination is quite fast and sometimes beans can germinate within a period of 14 days only. Also, with proper care and extra facilities, you can germinate beans within a week as well. As beans can grow the best under proper sunlight, it is always better to grow the same by using the process of hydroponics in a spot that can receive direct sunlight for the majority of the day. The concept of beans growing is, the more amount of light, the better will be the quality of the harvest.

In a hydroponics system, there are two types of beans that are grown most of the time.

Such beans are usually categorized depending on the habit of growing. The type of beans that usually grows in the low bush form are the ones that are known as the determinant variety, and the type of beans that comes with a habit of vining are known as the indeterminant variety. The indeterminant variety of beans needs extra support. If you are growing beans in a really small space, the bush type beans will be the perfect choice as the indeterminant type looks put for extra support and thus, you will need more amount of space for that.

Radish

Radish is known as the cool weather crop, and they can be grown quite fast. If you are looking out for a root crop that can be grown using the system of hydroponics in your home, radish a great choice. Radish can be grown right from the seeds. There might be some staggering cycles of growth but it is possible for you to harvest them all throughout the year. Also, radish can mix up very well with other types of vegetables and is a must to have in your kitchen. After planting the radish seeds, you can see the seedlings only within three to six days. There is no need for light as radish is a crop of cool weather.

Cucumber

Growing cucumbers hydroponically is very common both in greenhouses of commercial types as well as in homes. Cucumbers are a plant of vining type, and they can be grown very fast. There is no need to think about the variety of cucumber as any kind of cucumber is well suited to the system of hydroponics. Rather, you need to think about your purpose of growing and then select the type of cucumber, either cucumber or pickles. You can hydroponically grow cucumber from seeds and the speed of germination is quite fast. When provided with a temperature of 75 – 85 degrees, the germination of seeds takes about four to seven days. No matter whether you want to grow the European variety of cucumbers, the Lebanese ones, or the American variety, you can grow all of them by using a hydroponics system. Cucumber needs warm temperatures and high light for growing properly.

Spinach

Spinach is among those vegetables that can be consumed both raw and cooked. It is a perfect vegetable for growing in a medium that is based on water. The best part is that you don't need to put in any kind of extra effort for the conditions of lighting as spinach can also grow very well in conditions of low lighting. When a good growing environment is maintained along with proper climate, continuous harvesting of spinach is possible for up to a period of 12 weeks. Spinach is a leafy vegetable and comes along with several health benefits. After planting the seeds, you can see the sprouting of spinach from the third or fourth day.

Broccoli

Broccoli is somewhat like cauliflower, where the only difference is the color. It can be easily grown in a hydroponics system and most types of broccoli can be cultivated in this system. This green vegetable has adapted itself very well in the modern-day cuisine and is loved by all. It also comes with great nutritional value. Just like spinach, broccoli can also be grown from seeds. In case you have set up a small system of hydroponics, then you can start with sprouting broccoli or lead broccoli. It takes somewhat around 7 – 14 days for the germination of the broccoli seeds.

Strawberries

When it comes to the hydroponics system, fruit plants are most often considered to be very difficult to grow, but one of the exceptions is strawberries. While talking about the commercial systems of hydroponics, strawberries are among the most commonly grown plants. But, one drawback of strawberries in a hydroponics system is that they take some years for maturing right after planting the seeds. So, with strawberries in a hydroponic system, you need a little bit of patience. The medium of growth that you will be using for the strawberries needs to be well-drained and aerated as well.

Blueberries

When blueberries are grown in the traditional way in soil, they need acidic conditions for growing. But, when grown in a system of hydroponics, the requirements actually change. One of the primary problems of growing blueberries in the traditional way is that maintaining the level of pH is quite tough as they need an acidic environment. In a hydroponic system, it is quite easy to maintain the acidic level of pH. Blueberries are loaded with various types of vitamins, and consumption of the same is important for a healthy living. It is always recommended to use transplants for growing blueberries and not the seeds.

Adding to all the plants that have already been mentioned above, some other types of plants that can be grown in a hydroponic system are cauliflower, cabbage, mints, pepper, basil, beets, chives, celery, onions, peas, eggplants, yams, leeks, squash, and rhubarbs. Some other veggies that can be grown using the method of hydroponics are corn, summer squash, melons, pumpkins, and zucchini. But, the primary problem that comes while growing these vegetables is that they need a large space than normal, and that is the reason why they are not considered as being practical for growing in a system of hydroponics. You can grow them when you plenty amount of space for growing them vertically. Also, they are quite costly to be grown in a hydroponics system and the profits after harvesting cannot even compensate for the overall expenses.

So, the primary thing that you always need to consider while choosing plants for your hydroponics system is that they need to be compact, small, and much easy to be grown. The crops that are actually very easy to be grown in constrained spaces are bush varieties of tomatoes, beans, and cucumbers. Always keep in mind that although you are growing plants hydroponically, the needs of the plants would still be the same as it would be when grown in soil. The plants would still require light, heat, and water.

Also, you can grow root type vegetables in a system of hydroponics, but you need to choose a quite thick growing bed of the medium. It is not that they are nearly impossible to be cultivated but when you have a small amount of space, it would be wise of you if you do not choose them to grow. For instance, the majority of the people who are into indoor hydroponics prefer growing globe variety of carrots in place of the full-grown ones as they come with short length roots. Also, you can grow trees like bananas and trees in a hydroponic system, as long as they are provided with the perfect nutrition and lighting conditions.

If you want to grow flowers in a hydroponics system, you can do that. Even the types of flowers that generally would not have thrived in the climate of your place can be grown easily with the help of a hydroponics system. You have the choice of growing tropical as well as semi-tropical flowers too. So, if you are a lover of the unusual varieties of flowers and you want to grow them in your house, a hydroponic system would be the perfect choice for you. But, the only downside that comes with growing flowers in a hydroponic system is that you can only grow one type of flower, or you can even mix up flowers, but a system needs to have the same type of seeds. For instance, when you want to grow roses in a hydroponic system, the nutrient solution will be high in potassium that might not be suitable for the other plants.

If you are just starting with a hydroponic system, it is always recommended to start with herbs as they are low maintenance, and also they can be harvested quite fast.

Chapter 7: Plant Culturing

All the plants that are used for hydroponic culture start from the seeds. Some of the growers also opt for buying the ready to transplant type of transplants for growing their desired plants. But, when you decide to use transplants that have been grown by someone else, in some other place, it might have the chance of having a disease that can ruin the entire system. Let's have a look at some of the concepts of plant culturing.

Seeding

Grafting for growing plants is most widely used for plants such as peppers, tomatoes, and eggplants. In all those areas where there is high sunlight, such as deserts or tropical regions, you can transplant seedlings at a very early stage. In such types of locations, it will be beneficial for the growers if they can raise their own seedlings. For instance, cucumbers that are grown in the tropical regions can easily be transplanted within two weeks of sowing the seeds as the rate of growth of cucumber plants is quite high.

You can use various processes for sowing the plant seeds. You can directly sow the seeds into multiple trays using a mix of peat-lite or even perlite, rockwool, or vermiculite. There are various types of containers available in the local markets which can be used for sowing seeds, the most common one being the ice-cube trays.

Production of Seedling

The basic rule that needs to be followed while growing seedlings is, 'Good seedlings will result in good quality production.' The potential type of production for any kind of plant is actually set at an early phase in its cycle of life. A spindly and thin seedling will prohibit the proper flow of water along with nutrients as it has a thin form of stem base. Also, the overall production will get reduced considerably and there are high chances of the seedlings getting infected by insects and diseases. The seedlings might also develop the chances of falling over. This very principle is actually applicable to all types of plants.

At the time of germination, the embryo of the seed gets activated and grows by using up the food that is stored in the endosperm. Various types of biochemical processes happen for activating the seed embryo and for breaking down the starch or stored form of food for making it available for the growth of the embryo. For proper germination of the seeds, you will need to provide the required amount of oxygen, water, temperature, and also light. Also, sterilizing raw water, along with the proper application of chemical agents, can help in reducing any kind of developing diseases, but proper care needs to be taken for not damaging the tender plant seedlings. If you are using any kind of chemical agent, try to use them at lower rates.

Culture of Tomato Seedling

For tomato seedling, rockwool is often considered as the most ideal growing medium. Rockwool comes with great water retention capabilities, and is this perfect for tomato seedling culture. After you are done with sowing the seeds, try to keep them moist all the time and you are not required to cover up the seeds. Keep the medium under proper lighting conditions and temperature. But, make sure that the growing medium is not too moist or excessively dry as this might result in killing the seeds. Most of the commercial types of growers try to keep the seedlings in a special seedling zone of the greenhouse. Such a spot comes with overhead sprinklers that are controlled by using timers or sometimes moisture sensors. You can also the ebb and flow system in this case.

It is very important to note that each cycle of watering needs to soak the growing mediums properly. Uniform application of water should be targeted, or else uneven growth of plants will take place. Keep in mind that plants need less amount of water in dark conditions, when the weather is cloudy, and also during warm weather. As the plants tend to grow, their water requirement increases considerably.

One more thing that you need to take proper care of while opting for tomato seed culture is proper drainage. The solution of nutrients needs to pass along the growing surface and then drain away properly from the growing plants. This helps in proper leaching and also helps in moving the oxygen properly across the medium for proper germination of the seeds. This also helps in the subsequent formation of roots. You can use mesh benches that can provide you with proper drainage and will also allow pruning of roots by air so that the roots of the seedlings can turn out to be denser in nature.

You need to maintain a proper temperature of 79 degrees Fahrenheit to 82 degrees Fahrenheit during the initial phase of germination. As the seedlings tend to grow, the cotyledon of the seed expands. Try to maintain the day temperature at 73 degrees Fahrenheit and the night temperature to 67 degrees Fahrenheit and keep in regulating this temperature setting for many weeks.

After the seedlings have evolved, you can start with a daily program of spraying for preventing diseases, and it will also be helping the plants to grow healthy and for coping up with the stress of transplanting.

Culture of Cucumber Seedling

Similar to tomato seedlings, cucumber seedlings can also be grown in cubes of rockwool. But, as the germination rate of cucumber is much more than tomatoes, you can sow them directly into the blocks of rockwool. You are required to presoak the blocks before sowing, and proper procedures of watering also need to be followed. This will help in preventing the seeds from getting dried. Also, in all those areas where the solar radiation is very high, you can cover the seeds in the cube holes with some vermiculite for preventing the seeds from getting dried up. During the early stages of germination, prevent the use of fungicides as it might result in slowing down the overall process of germination and might also cause stunting of the cucumber seedlings. You can start using hydrogen peroxide at 30 ppm for the seedlings and maximum up to 50 ppm in case any algae are present.

While seed germination, try to maintain the air temperature at 74 degrees Fahrenheit. The seedlings will germinate within two days. Also, the block temperature should not exceed 73 degrees Fahrenheit for the next seventeen days of germination. After that, the air temperature should be 79 degrees Fahrenheit at the day and 70 degrees Fahrenheit at night.

In the northerly latitudes, try to carbon dioxide level between 7500 – 800 ppm. Right after the leaves start to overlap, within a period of ten days, start spacing the blocks in the way of checkerboard for reducing the density of the cultivation by half. Also, as you increase the space, this space will prohibit the plants from getting too tall along with being spindly, which is a very common condition for diseases and infections. You might need to re-space the seedlings again if the leaves tend to overlap once more if you are holding them for more than three weeks.

If you are growing the seedlings on benches of wire mesh, the roots of the same will tend to be air pruned as they will be growing out from the blocks. This will be forcing the plants to generate more roots right within the blocks and also reduces the shock of transplanting if you are going to transplant them. In case you have decided to transplant the seedlings, try to lower down the house temperature of the seedlings to 70 degrees Fahrenheit at day and 68 degrees Fahrenheit at night. Try doing this before several days of transplanting. In all those areas where the sunlight the very harsh, you can transplant the seedlings at earlier stages.

Culture of Pepper Seedling

Just like tomatoes, germination of pepper seed might vary, and so you are needed to figure out the extra amount of seeds that you might need by utilizing the percentage germination that comes along with the seed package. Try adding an extra 10-15% as vigor and germination might not be uniform. Many growers of pepper utilize round plug type trays for sowing the seeds of pepper into the granular type of rockwool. This permits them to choose the most vigorous and uniform seedlings for the purpose of transplanting to the blocks of rockwool. You can use cubes of rockwool for sowing at the initial stage. After about 15 – 18 days, the peppers sowed in the cubes of rockwool can be double spaced by cutting down the cubes in strips for making two units of flats from each.

While germinating the seeds of pepper, try to maintain the temperature between 78 degrees Fahrenheit and 80 degrees Fahrenheit during the initial stages. The RH or relative humidity needs to be 80% - 85%. Try maintaining these similar temperatures during the night as well. After the seedlings properly emerge, lower down the air temperature to 70 degrees Fahrenheit – 72 degrees Fahrenheit. You need to provide the same with a supplementary form of lighting for almost 16 hours a day. After about 3-4 days of seedling emergence, decrease the RH to 65 – 70%.

Generally, after 18 – 20 days of seed sowing, you can transplant the seedlings to blocks of rockwool of 4 inches, right after the appearance of the first true leaves. The plants will need to be supported with sticks if you permit them to grow for more than four weeks.

Culture of Eggplant Seedling

The overall procedure for sowing the seeds, temperature, requirements of light, and CO_2 is more or less like that of tomato seedlings. After about 18 – 20 days, you need to transplant the seedlings to blocks of rockwool, just like peppers.

Culture of Lettuce Seedling

For both the NFT and raft culture, the best way of seed sowing in one-inch cubes of rockwool. You need to presoak the rockwool cubes by using raw water, just like peppers and tomatoes. If you are using automatic seeds, you can opt for pelletized seeds. It will result in uniform germination more than raw seeds as it comes with clay coating that can successfully retain all the moisture right around the area of the seed at the time of germination.

During the initial stages of germination, the temperature cannot exceed 73 degrees Fahrenheit. Temperature more than that might result in inducing dormancy. If there is a problem of dormancy, try maintaining the temperature at 61 degrees Fahrenheit until there is the emergence of the seedlings. You need to maintain the RH at 65% - 80% and the level of carbon dioxide at 1000 ppm during day time. You also need to use a supplementary form of lighting for 24 hours at the time of emerging seedlings during the months of winter in the northern areas.

Chapter 8: Pest Control

When you are setting up a hydroponics system for gardening, your system will not have that degree of risk of infestation from pests as you would have normally when using traditional cultivation using soil. But, there are still certain concerns regarding the protection of the plants from specific pests. There is nothing wrong with being vigilant by nature when you are developing a hydroponics system as it naturally needs some extra care. It is always better to prevent a problem of pest much before it happens, actually. So, in this chapter, you will be learning about the various aspects of controlling pests in a hydroponics system.

Common Types of Pest Problems

Right before controlling pests, you are required to learn about the various types of pests that are common in hydroponics systems. Let's have a look at them before jumping in for the remedies.

Aphids

Most of us have learned about this type of pest during our school days, and you must have thought that you are free from them. But, unluckily, they attack systems of hydroponics, especially when the plants that you are growing in the system have an excessive amount of nitrogen in the source of food. This type of pest is generally found across the stems of plants and they might either be green, black, or tan in color.

Whiteflies

This type of pest might turn out to be very tricky, but spotting them is not that tough. They look similar to the very little white-colored moths and they tend to fly away as you try to catch them.

Spider Mites

They are even much smaller than whiteflies and are generally less than 1mm in length. This type of pest is often regarded as the most dreaded type of infestation when talking about a hydroponics system. They look somewhat like tiny spiders. But, because of their excessively small size, they can run away easily and might not come under notice until it actually gets out of your hands.

Fungus Gnats

This is another type of tricky pest as the grown type of gnats is not that harmful to the plants, but the larva of the same is. You can notice the pest larvae eating away at the root portion of your plants, and this might result in the bacterial form of infections very easily and quickly.

Thrips

Thrips are like aphids and comes with the capability of turning the plant leaves brown or yellow as they feed on the nutrients and generally sucks out the same. They are slightly bigger in size, somewhat around 5 mm, but they are still pretty hard to spot. They generally appear as black dots right on the top portion of the plant leaves.

Avoiding Things That Might Invite In Pests Into the Growing Area

There are some good types of practices that you can adopt to reduce the overall chances of pest infestation in the hydroponic system. Let's have a look at them.

Preventing Going In Dirty Condition

Right before you enter the growing area, you are required to be sterile by wearing clean clothes. Various types of pests, bacteria, and other types of contaminants might cling to the clothing if you fail to notice them. Even if you feel that there is nothing, there is no good in risking it. If you fail to do so, you might invite in pests to the growing area that can actually destroy your entire system. Also, you need to take proper care of the tools and equipment that you generally bring in to the growing area.

Giving the System Sterile Start

If you are just setting up a new system or even doing some sort of work on the same, it is always better to be sterile. You need to properly clean the tanks, vents, fixtures, or any type of gear before introducing the same to the growing area. Also, you need to check the seal quality around the growing area. While you might be looking for a well ventilated growing area, you also do not want a free entry for the pests. So, check the seals from time to time.

Outside Tools

Pests have the capability of cropping up from any sort of sneaky place and also along with the tools and materials that you introduce in the garden. So, every time you bring in something new in the growing area, sterilize the same properly and store them in a proper place which is clean and tidy.

Growing Media

You might think that this is something to generate scare campaign as in most of the cases, the growing mediums are safe and sterile. But, there are still certain things that you need to look out at. In case you are using an organic medium, such as rice husk or coco coir, pay proper attention. This type of medium can easily harbor pests, and thus, they need a special type of treatment. You need to ensure that the growing media that you are going to use is properly sterilized and if possible, try to treat them for eliminating pests, if any.

New Transplants

At times when you are introducing new transplants to the system, you need to be extra careful. The plants that are brought from outside can easily carry in fungi, bacteria, pests, and diseases. For dealing with all these risks, you are required to get the new transplants from a place that is well-maintained and clean. Also, right before transplanting anything, take your time to examine them properly for any kind of health problem.

Watching the Humidity

Pest control needs to be practiced from the very first day of setting up your hydroponics system. The first line of defense against pests will be to put in the measures that can deal with them. Some types of pests, such as fungus gnats and spider mites, are attracted by excessive moisture and low humidity. If you try to keep the humidity, not below 50%, you can prevent any form of pest infestation. Also, you need to make sure that the moisture content of the system, along with the growing area, is not much as it can easily invite in various types of pests for residing in the system.

How Can You Identify Pest Problems?

Even with the most diligent type of prevention, you can still have pest infestation into your hydroponics system. Like any type of setup for hydroponics, you are required to examine the plants daily for finding out problems, if any. Also, you should not be mixing pest signs with other types of issues, like disease or nutrient deficiency.

Discoloration

Whenever pests suck out nutrients from leaves, such as aphids, you can notice discoloration of the leaves and they often turn yellow in color. This very discoloration is centered by the tiny form of holes from which the infested pests feed on and not just spread around on the leaves.

Spots

There are certain pests that can leave back a signature spot pattern, whether yellow, white, black, or brown. If you come to see any kind of spots on the leaves, immediately check if there is any form of deposits on the plant leaves, such as feces, eggs, etc. If the spots on the leaves tend to be scraped off, there is a definite issue of pest in the system. If you come to notice this type of spot on any of the plants, make sure that you check other plants in the system as well to make sure of the degree of infestation and also the type of pest.

Pest Holes Vs. Lesions And Burns

Whenever you come across any kind of hole on the leaves, it is very easy to make your own assumptions. That is the reason why it is so important to have a closer look and properly check the edges of the holes. Burns can be determined very easily as they tend to appear in all those places that are close to heat and light sources and will show discolorations or burns around the holes. The pests that most often infest on any type of hydroponics garden are more like being suckers than munchers. This indicates that the holes left behind by the pests are generally tiny and are often surrounded and raised by a whitish or yellowish area.

Things to Do When Your System Has Pest Infestation

If you have come across any of the above-mentioned symptoms in your plants, there is a problem of pests, and you are required to get that fixed as quickly as possible. But, when any pest has already succeeded in getting in, it might actually be a crucial job to mitigate the very issue properly. Pests can spread in any hydroponic system very fast, so as soon as one plant gets affected, the others are most likely to get the same thing quickly.

Not Waiting Any Longer to Take Care of the Pests

When you act immediately after you have spotted pests in the system, you might be capable of sparing the unaffected plants. But, if you decide to wait, you will most likely be ending up with a completely infested system.

Determining the Intervention Level That Is Needed

Some of the pests can be taken proper care of by altering the environmental conditions, removing them manually, or by using other alternatives, while there are certain pests that can only be dealt with chemicals. But, unless and until you are absolutely sure about pest infestation, try not to use any kind of chemicals in the system.

Sticky Traps

The most common way of dealing with a pest infestation in a hydroponics system is by using sticky traps. They work the same as the other types of sticky traps for bugs and might turn out to be very helpful if you are dealing with pests that have shorter cycles of life. Also, while using sticky traps, you can easily determine the type of pest and identify the same. If you can successfully identify the pests, you can opt for a better route in order to get rid of the pests. Also, if your plants are not infested by pests, it is helpful to keep sticky traps as a measure of prevention. If you find out any pest caught in them, you can easily take the necessary steps for preventing any larger kind of issue.

Natural Solutions

There are various types of solutions that are available in the market, but you will not be taking any chance of killing the plants in the system. When you are in confusion or doubt, ensure that the solution you are going to use comes along with guaranteed plant safety. You can completely rely on natural solutions such as Pyrethrin. It is extracted from natural substances such as chrysanthemums and can easily prevent pest infestation.

Information On Pests and Treatments

Let's have a look at the treatments of some of the most common types of pests.

Aphids

This pest tends to secrete a sticky kind of residue that can easily stimulate the development of sooty molds. They have the tendency of sucking out nutrients from the plant leaves. One can easily spot the aphids moving around the stem area, although their color might vary. You can start using predator bugs that can eat up aphids for controlling the infestation. The most common choices of bugs are lacewings and ladybugs. Stems, leaves, or even the plants that are badly infested by the pest might need to be completely removed.

Fungus Gnats

They are not much harmful, but the problem might be created by the larvae of the same. They mostly feed near the root area of plants. They can be seen flying way whenever an infested area is disturbed or touched. The plants that get infested by them are most likely to turn yellow and might look ill. They can be treated by not using excessive water for the plants. But, if you have already overwatered the plants, allow the medium to dry out as much as possible. The eggs can be caught by using sticky traps and you can also use nematodes for taking proper care of the larvae. You can also opt for spraying neem oil if the infestation tends to go out of control.

Spider Mites

They leave back fine webs all over the infected areas of the plants and might turn out to be a very crucial pest. As they suck on the leaves, you can notice whitish and yellowish spots on the infected leaves. They can very quickly grow in number. So, try to check beneath the leaves where they tend to gather. You need to manually remove the areas that are infested badly by pruning and then removing the infected areas. You can use a safe, natural insecticide for bringing them under control. You can spray a mixture of wetting agent and neem oil every week for killing the eggs.

Thrips

They can grow to a huge population within a very short amount of time and might result in a dangerous form of infestation if left without any kind of treatment. They leave back black spots on the leaves. You can use insects that feed on them for killing them. Ladybugs and lacewings can be used. You can also use minute pirate bugs for the most effective results. If the infestation goes out of your hands, you can use pyrethrin.

Whiteflies

They tend to hide under the leaves and might look like minute moths. They leave back sticky residue on the leaves. For getting rid of them, you need to spray the infected plants with water at low pressure. You can also introduce bugs for eating them up. For a quick remedy, spray neem oil along with organic insecticides.

Chapter 9: Extra Tips and Tricks for Beginners

Hydroponics gardening is gaining its popularity slowly, and with its craze among amateurs, it has reached new heights of popularity. Hydroponics comes along with several factors that are making it so popular, much more than the traditional way of cultivation, such as you can grow a large number of plants that too in a very constrained area, need less amount of water for the plants, controlled lighting systems, higher yields, easy harvesting, and high-quality plants. All of these tend to make hydroponics system a perfect choice for all the beginners. Also, there is almost negligible usage of pesticides and is much healthier than the conventional way of gardening using soil. It is cost-effective overall and has been developed for improving the overall productivity with the least amount of cost.

Also, in a hydroponics system, as you have learned in the earlier chapters, the plants can grow 20% faster when compared to the traditional systems. Hydroponics comes along with a wide array of benefits, but right before starting ideal large-sized cultivation based on a hydroponic system, it is always better to set your hands on and experiment with the system by setting up a smaller system, especially when you are only a beginner. So, in this chapter, you will be coming across some important tips and tricks that can help you gain all the knowledge about a hydroponics system.

Planning the Location

No matter if you want to set up a large commercial system of cultivation or just a small one in your home, location always plays a very important role in the system of hydroponics. Location is not only about the involvement of growing your plants outdoors or indoors but also for determining whether you will be growing the plants in a tent, box, or room. In case you just want to start with a small room, it would be great if you could paint the walls white or cover them by using Mylar as it will help in reflecting back the lights to the surface of the plants. Also, planning is not only for the location but for the whole setup. Try to fix in your mind all the things that you want to achieve and then proceed with the same.

Choosing the Plant Types That You Want to Grow

The majority of the plants can be grown in a hydroponic system.

But, if you are just starting with the system, it is always better that you opt for the small-sized plants. Try choosing all those plants that come with very less requirement of maintenance along with nutrients. As a beginner, you can opt for plants such as veggies and herbs. Also, you can opt for the fast-growing plants as it will be much easier for you to assess the performance of the system that you have set up. Adding to that, try to opt for grow plants of the same type at the same time if you are completely new to the system as they come with the same requirement of nutrients. So, if you are a beginner and you want to learn some techniques involved with a hydroponics system and how to grow plants in the same, plants that come with low maintenance would be the perfect choice for you.

Making a Proper Plan

Right after you have set your mind regarding what crops you would like to grow in your hydroponics system, now it is the time to chalk out a proper plan. By making a plan, I mean to know all the nutrient requirements, the lighting periods, and also the requirements that you would need. Having the perfect supplies for an indoor hydroponic garden is of utter importance. Try making out a list of all the things that you would need for running the system without any kind of problem.

Choosing the Ideal System of Hydroponics

When it comes to hydroponics, there are several methods or systems that are available today. All the methods come along with its own set of advantages and disadvantages. The proper method of hydroponics needs to be selected depending upon the complexity and also the scale of your farming. As a beginner, the perfect system would be the deep water cultivation as the setting up process of the same is super easy and is economical in nature as well. If you are willing to set up a large system, nutrient film technique would be a great choice for you as it uses up less amount of electricity, water, and nutrients. Along with all these, there are several other methods that are available for you, and you need to select the perfect one for you depending on your requirements. As a beginner, do not make the mistake of choosing any kind of system without even considering the plans as you might end up with zero results. So, always select the system after you have a proper plan.

Take Care of the Lighting Conditions

When it comes to edible types of plants, they need a minimum of six hours of light to a maximum of sixteen hours per day. If you have opted for the indoor type of hydroponic system, it is very important to take proper care of the lighting conditions as it is essential for the plants to grow and flourish properly. Generally, every plant needs around 50 – 70 watts of light. Also, if you are going for the readymade hydroponic system, they come with an inbuilt system of lighting. If you are setting up the system on your own, opt for HID bulbs or LEDs. You need to take care the source of light is able to produce about 6000 – 10,000 lumens of light for every square foot. Light is really important for proper growth of the plants as only with proper nutrients and care; the plants will not be flourishing.

Maintaining Proper Temperature

As a beginner, most people opt for the indoor system of hydroponics. In that case, maintaining an ideal temperature for the growing room is essential for the growth of the plants. Try to control the temperature as per the requirements of the plants. You need to note that the growth of plants tends to stop whenever the air temperature goes above 85 degrees. So, you are required to keep a proper check on your indoor garden. Also, the temperature of the growing medium needs to be moderate. Remember that the air temperature of the growing area is most likely to increase because of the lighting systems. So, in that case, you can use a mild heater when the climate is cold and maintain proper ventilation by using fans when the temperature tends to be very high.

Germination of Seeds

Right before starting with hydroponics gardening, another essential thing that you need to do is to spur seed germination that you have decided of planting in the system. You can do this by using traditional systems. For this, start by moistening a paper towel and then lay down the seeds carefully on the surface of the moist towel. Try t cover the seeds with a flat surface. For spurring germination of the seeds, place them in a dark spot for about 24 hours.

Planting Process

After you are done with the process of seed germination, you will have to plant the seeds in your preferred medium type. As you use a medium for growing plants, make sure that you have submerged almost 70% of the medium in the solution. Also, positioning the medium is very important. For the best growth rate, you need to ensure that the medium sits exactly above the solution surface.

Checking Water Quality

Make sure that you always use soft water instead of hard water. Hard water comes along with high content of minerals that might prevent the nutrients from getting dissolved in the same properly. Also, you are needed to keep an eye on the level of pH. Try to maintain the level between 5.9 and 6.7.

Selecting Proper Nutrients

You are required to have a proper idea about the nutritional needs of the plants that you have decided to grow in the system. You need to have detailed knowledge regarding the quantity of nutrients that are needed by the plants every day and weekly. Also, while using fertilizers and insecticides for the plants, make sure that they are free from chemical content and are also safe for the plants. You need to ensure that the nutrient percentage that you will be using in the solution matches the nutrient requirement of that of the plants but note that the percentage of nutrients in the solution cannot go above the required percentage. So, it is always better to use less than the required percentage and keep on increasing the same from time to time in small quantities, depending on the performance of the plants.

Checking Health of Plant Roots From Time to Time

The health of roots is very important. It is the one that will be providing the plant parts with the required nutrients from the solution. So, maintaining the health of the same is very important. Try to check the health of the plant roots from time to time in order to prevent any form of damage to the plants. In the roots get damaged, it won't be able to take up the nutrients to the plant parts. Try to minimize the exposure of light from the solution for avoiding any form of growth of fungus and algae that might lead to root damage. Also, pests are very much prone to the roots and try to feed on the same most of the time. Opt for regular checking of the same for safeguarding the health of the plants.

Making the Required Investment

After you have selected the type of plants that you want to grow in the system, it is very important to maintain the same in the proper way. Do not just opt for the cheap option of tools such as cheap pumps, trays, etc. as it might ruin the entire system. There is no need for starting largely from the very beginning. Start with a small system and try to expand the same slowly. But, make sure that you use the perfect types of equipment and do not just look out for the cheap options.

Chapter 10: Common Mistakes and How to Avoid Them

People enter the hydroponics world for various reasons. It can either be for profit, for fun, or for both. Like any other sector of gardening, the more you will tend to do, the more amount of knowledge you will acquire. The better is your knowledge, the better you can deal with the complexities and adversities that come up in the aspect of soilless cultivation. As you start with it, hydroponics needs a lot of planning and knowledge, and by doing these, you will be able to prevent yourself from making some of the costly and time-consuming mistakes. In this chapter, you will be learning some of the most common mistakes that are made by the beginners and also by some of the experienced growers.

But, what is the need to focus on the mistakes? Well, in hydroponics, there is a curve of learning that many people try to omit and opt for shortcuts. They end up with zero results and being a complete failure. We will be focusing on the common types of mistakes as we are human beings and we all tend to learn more from the mistakes that we make. With proper accounting of these mistakes, you will also be able to improve the features of your hydroponics system. So, let's have a look at them.

Underestimating the System Building Costs

As a home grower, a system of hydroponics can be set up with as little or as much you actually want to spend on the system. But, underestimating all these costs regardless of the size of the system, it can actually go out of hands along with a system that you will not be able to use properly. Hydroponics comes along with various types of systems, and different systems come along with different setup costs. There are certain systems that can even be built without purchasing anything from the market and using household items. Right before you set up your system, it is better to jot down the requirements in the first place and then set your budget accordingly for preventing any kind of future problem.

Selecting the Wrong Type of Crops

Thinking that all types of crops can grow in all types of hydroponics systems is one of the biggest mistakes that is made by most of the growers. Every plant comes with its individual needs. Also, some plants are only suitable for certain environments. Growing outside in your greenhouse or indoors or in any other type of growing space comes with distinct importance, but before that, there are three questions that you need to ask yourself right before buying plant seeds that you want to grow.

- What are the techniques that you are going to use?
- Is there any kind of weather constraints?
- Is it possible to grow your desired plants using your technique of production?

All plants come along with different needs of themselves. There are short as well as tall plants, and all can be cultured in certain specific ways. Suppose, if you have set up a wick system and want to grow tomatoes in it, you will fail. Also, climate plays a very important role in the growth of plants using hydroponics. It is true that you can regulate the temperature in a hydroponic system, but in places where the heat is excessive, you will have zero chances of cultivating crops of cool weather, and vice versa. With an excellent temperature controlling system, it is possible to grow crops of any climate type, but the overall cost will also increase.

Not Paying Attention to the pH Levels

pH level is one such thing that can easily destroy a hydroponics system. The mistake of not paying attention to the levels of pH starts when the growers want to see faster results and starts mixing up different nutrients of their wish.

The excessive urge to see the results prevents the growers from considering the formulas along with the attached effects of the same. When the levels of pH are completely out of balance, the plants will suffer and will ultimately end up dying. The levels of pH help in determining whether the nutrient solution is alkaline or acidic in nature. The neutral level of pH is 7.0 in which most plants prefer to grow.

Also, when there is a problem with the pH level, various diseases related to the plants might arise along with nutrient deficiencies. So, it is very important to maintain a proper pH level in the system of hydroponics in order to facilitate the proper growth of plants. You can use a testing kit of pH level along with adjusting compounds for the proper pH level.

Using Wrong Nutrients

Most of the growers who just started with hydroponics fail to understand the proper nutrient requirement for the plants. Also, mixing the nutrients in proper proportions is very necessary to ensure that the plants get all that they need. Adding up too many nutrients in the solution is a very easy thing to do, and it will end up burning down the nutrients. It will be impacting the growth of plants from that very point of time. For overcoming this type of problem, you need to first learn about the nutrient requirement of the plants and then follow the feeding schedule that comes along with the nutrients.

For example, if the nutrient comes with the guide of using 2 spoons of the nutrients for the solution, try using only half the spoon. This will help in maintaining the level of pH in range. If there is any form of nutrient deficiency, you can increase the nutrient content accordingly, but not too much at a time.

Overwatering

The majority of the growers have the notion that plants need water and sun every day. This ultimately results in overwatering the plants. This is a very common mistake that can be coupled with the failure of a hydroponics system. As you overwater plants, the plants might droop and in certain cases, it might even lead to rotting of roots and plant death. It is very important to maintain the proper watering of plants in the hydroponics system. You can make certain adjustments for restoring the health of the plants.

One of the easiest ways of testing your schedule of watering is by testing the upper inch of the medium of growth. For example, if you are using coco coir when you press your finger on the top of the medium, and it pulls out dry, it is the sign of no moisture. Now, it is the perfect time for watering the plants. While using a pump for your hydroponics system, it will take some time for testing and finding out the errors for properly balancing the moisture level, relying on the type of system that you have. As you overwater a plant, you are actually depriving the plant of proper oxygenation. So, you need to balance the watering schedule according to the system for the proper growth of the plants.

Insufficient Lighting

There are many growers who do not give proper attention to the lighting setup. Light plays a very important role in a hydroponic system. If you tend to use the wrong kind of lights, the plants will not grow. Also, if you opt for the cheapest options, they might not function in the required way. The type of lighting that is suitable for the system types have been discussed in the previous chapters. You need to adjust the light intensity and also the quantity of the same according to the size of your system. Insufficient lighting might result in poor growth of the plants and even death.

Conclusion

Thank you for making it through to the end of *Hydroponics for Beginners: A Simple Introduction And Guide To Gardening Without Using Soil.* Let's hope it was informative and able to provide you with all of the tools you need to achieve your goals whatever they may be.

Now, your only job is to plan out a way in which you can set up a hydroponics system. Trying using up all the tips and tricks that you have learned from this book and make good use of the same.

If you are a find lover of gardening, hydroponics is the perfect choice for you as it will be providing you with more control over your garden. Just focus on your objective and monitor the system daily for the best results. Hydroponics is not a tough game and you can adapt to the system very easily. All that you need is a bit of knowledge about the various types of systems, along with the requirements of the plants. Your only task is to provide the plants with the required nutrients and take proper care of the same.

The best thing that you can do is to first gain some knowledge about the types of plants that you would like to grow in your own hydroponics system and then move ahead with the other steps. Try to keep the plants free from all types of pests and diseases and prevent the use of chemicals as much as you can.

Finally, if you found this book useful in any way, a review on Amazon is always appreciated!